ENTROPY ANALYSIS IN THERMAL ENGINEERING SYSTEMS

ENTROPY ANALYSIS IN THERMAL ENGINEERING SYSTEMS

YOUSEF HASELI

School of Engineering and Technology
Central Michigan University
Mt Pleasant, MI, USA

ELSEVIER

ACADEMIC PRESS
An imprint of Elsevier

Academic Press is an imprint of Elsevier
125 London Wall, London EC2Y 5AS, United Kingdom
525 B Street, Suite 1650, San Diego, CA 92101, United States
50 Hampshire Street, 5th Floor, Cambridge, MA 02139, United States
The Boulevard, Langford Lane, Kidlington, Oxford OX5 1GB, United Kingdom

Library of Congress Cataloging-in-Publication Data
A catalog record for this book is available from the Library of Congress

British Library Cataloguing-in-Publication Data
A catalogue record for this book is available from the British Library

ISBN: 978-0-12-819168-2

For information on all Academic Press publications
visit our website at https://www.elsevier.com/books-and-journals

Publisher: Lisa Reading
Acquisition Editor: Maria Convey
Editorial Project Manager: Joanna Collett
Production Project Manager: Selvaraj Raviraj
Cover Designer: Mark Rogers

Typeset by SPi Global, India

Dedication

Dedicated to the memory of my Father,
Tahmaseb Haseli

Contents

Preface

It was about two centuries ago that Nicolas Léonard Sadi Carnot, a French military engineer, presented an influential treatise. Although remained unappreciated for a decade, it provided a profound basis for investigations of his successors and the advancement of the Science of Thermodynamics. Carnot's research on the theory of heat engines was itself founded upon the caloric theory, empirical findings of his predecessors, and philosophical reasoning. The invalidity of the notion of heat as an indestructible matter had become obvious among the pioneers by the mid-19th century. There were compelling experimental evidences supporting the equivalence of heat and work, the first main principle of the Mechanical Theory of Heat, according to which heat can be produced by expenditure of work and vice versa.

Unlike the first main principle whose statement and formulation can readily be understood by a student of an average intelligence, concepts like entropy originated from the second main principle of the Mechanical Theory of Heat appear to be challenging, perhaps, for everyone who has undertaken an introductory class on the subject. Such concepts are invented through a formulation of the second law of Thermodynamics. However, the analytical formulation of the second law is not a mere expression of the experimental observations—that heat cannot be converted completely into work, or heat cannot spontaneously transfer from a cooler to a warmer body. It involves a hypothetical concept, reversibility, which may only be realized in an imaginary process; which may be regarded as a preliminary source of difficulty in understanding the entropy-related concepts.

Today, after over 150 years of invention of entropy by Clausius, still there remain confusions surrounding the concept of entropy and the phenomenon of entropy increase. One may find a variety of interpretations or descriptions for entropy such as arrow of time, measure of disorder, chaos, wastefulness, and energy dispersal. On the other hand, some argue that understanding of entropy is only possible through statistical mechanics. The first question may cross a curious mind is: Why is not there a universally agreed interpretation for entropy yet? It has a simple definition $dS = dQ/T$, a differential of S (entropy) is equal to the differential of Q (heat) divided by T (temperature of body). The explanation given by Clausius as the inventor of entropy is that S represents the transformational content of a body like U that denotes its (internal) energy content. All we know nowadays about Clausius is his inequality with no adequate mention of what he meant by entropy and how he discovered it.

Despite entropy remains as a gray area (it is not as clear as many other concepts deduced from natural laws), today entropy-based analysis has frequently been employed as a design tool in a wide range of applications. Often, second law-based studies present entropy calculations but without any constructive use of such calculations. It is natural to ask: What is the goal of entropy-related calculations? Is entropy generation always an indication of losses, for instance, in a power cycle, fluidized bed, boiler, hydrogen production plant, chemical reaction, condenser? Do we need to be always concerned about the growth of entropy? Are there specific circumstances where entropy-related calculations may yield meaningful results? The primary objective of this book is to highlight the limitations of the application of entropy in engineering and clarify when a second law analysis may lead to rewarding results.

The journey of the present book begins with an overview of the fundamental thermodynamic concepts in the opening chapter. It is then followed by a brief historical sketch of Thermodynamics in Chapter 2, which illuminates its evolution as well as the contributions of many ingenious men to the advancement of the subject during the 19th century. More importantly, a careful examination of several sources reveals that the tutorial method of the second law and entropy could be much easier had it followed the same path as it was discovered and presented by the founders. The current method of teaching the second law, inherited not from the original founders but those authors who developed first textbooks on Thermodynamics in the late 19th and early 20th century, skips important steps, for instance the role of the ideal gas law in the investigation of Carnot, Thomson, and Clausius.

A detailed discussion on the shortcomings of the common tutorial method of entropy is presented in Chapter 3. Specifically, the demonstration technique of the Carnot's corollaries that rests on philosophical reasoning is shown to suffer from certain issues. The common derivation method of Carnot efficiency and introduction of the absolute temperature scale without a proper background is critically reviewed. A simple but effective method is then proposed to ease understanding the connection between the chain of concepts like Carnot efficiency, entropy, reversibility, and absolute temperature. The discussion will advance in Chapter 4 where the main task is to clarify the phenomenon of entropy increase and to show the direct connection between the phenomena of heat transfer and entropy generation.

Chapter 5 presents a comparative assessment of the efficiency of common heat engines. The chief goal is to illuminate that determination of the most efficient engine is contingent on specific assumptions. For example, the Carnot engine along with the Stirling and Ericsson engines are said to possess the highest efficiency among all heat engines subject to an assumption that the highest and the lowest temperatures are the same for all the engines. If, however, the engines are constrained to experience the same degree of compression, the Carnot engine is no longer the most efficient design.

Our investigation continues by applying entropy analysis to simple and advanced power cycles. The objective is to show that entropy production may become equivalent to an efficiency loss under specific conditions. We will see in Chapter 6 that in endoreversible heat engines, a class of theoretical heat engines which experience external irreversibility only, the thermal efficiency happens to inversely correlate with the entropy production. Nevertheless, in practice, engines do also experience internal irreversibilities. It will be shown in Chapter 7 that a design based on minimum entropy production rate in irreversible engines operating in closed cycles is not equivalent to either of maximum power and maximum efficiency designs. The three designs may, however, become identical if, for instance, the thermal energy supplied to an irreversible engine operating between a heat source and a heat sink, or the power output is treated as a fixed parameter.

In Chapter 8, we investigate the applicability of a second law-based analysis in conventional thermal power plants such as gas turbine and combined gas/steam cycles, which are usually driven by fuel combustion. In this chapter, the concept of *specific entropy generation* (SEG) is introduced, a new parameter that measures the entropy production of a power cycle per unit of fuel burned. It will be shown that SEG unconditionally correlates with the inverse of the cycle efficiency, and it can be viewed as a measure of efficiency losses in combustion-driven power generating systems. An application of the SEG concept to typical thermal power plants is explored.

An investigation on the application of entropy analysis to fuel cells is presented in Chapter 9. The primary objective is to show that the theoretical efficiency of a fuel cell is not bound by the efficiency of a Carnot cycle operating between the same low and high temperatures. Chapter 10 examines possibility of any connection between entropy and chemical equilibrium. A careful assessment of the Gibbs criterion of equilibrium reveals that the characterization of a chemical equilibrium by minimum Gibbs function is simply a postulation without a strong experimental evidence or theoretical proof. The last chapter explains the exergy concept and describes how it is originated by combining the first and the second laws. It is shown that a conclusion drawn from an exergy analysis may often be obtained from an entropy analysis.

It is hoped that this book will help readers to improve their knowledge and comprehension of the second law-related concepts and to have a clearer understanding of the applicability area of entropy-based analysis.

Yousef Haseli
Michigan, United States

Acknowledgments

I wish to express my appreciation to the American Society of Mechanical Engineering (ASME) for permission to use my articles published in the *Journal of Energy Resources Technology* Vol. 140, paper No. 032002, Vol. 141, paper No. 014501, and the *Proceedings of IMECE 2018*, paper No. 86510 in this book. I am deeply grateful to the staff at Elsevier, Maria Convey (acquisition editor), Joanna Collett (editorial project manager), and Selvaraj Raviraj (production project manager) for their assistance and guidance throughout the preparation of this book.

Over the years, I have had useful and joyful discussions with a number of people. I would like to thank, in particular, Professor Peter Salamon of San Diego State University, Professor Bjarne Andresen from the University of Copenhagen, Dr. Katherine Hornbostel of Pittsburgh University, and Professor Brian Elmegaard of Technical University of Denmark. I am eternally grateful to Professor Adrian Bejan whose pioneering work and influential articles on the subject have been inspiring.

I would like to acknowledge with gratitude the research support of Central Michigan University over the past two years.

Finally, I am indebted to my family for being patient and understanding while I devoted time to preparation of this book.

Yousef Haseli

CHAPTER ONE

Fundamental concepts

1.1 Thermodynamic properties

In many branches of science, property refers to the condition and characteristic of a substance under study. The properties of matter are categorized as mechanical, physical, thermal, etc. The examples include elasticity, yield strength, hardness (mechanical properties); density, melting point, viscosity (physical properties); thermal conductivity, thermal diffusivity, specific heat (thermal properties).

In thermodynamics, properties describe the state or condition of a substance. The basic properties frequently used in thermodynamic calculations are temperature, pressure, specific enthalpy, specific internal energy, specific volume, and specific entropy. The thermodynamic properties depend solely on the state of a given system or substance. They are independent of the path or process through which the system is brought to that state.

Some of these properties can directly be measured like temperature and pressure, whereas some properties are unmeasurable such as specific internal energy and entropy, which are determined using the measurable ones. Furthermore, the unmeasurable properties of two different substances may differ at identical pressure and temperature. For example, the specific enthalpy of air at 1 atm and 298 K is different from the specific enthalpy of water at the same pressure and temperature.

The properties of a wide range of common substances can nowadays be found in standard textbooks, web-based software, and commercial software packages. In thermodynamics, the state of system is said to be fixed if two independent properties are known. The other properties can then be determined at the given state with the use of a software or thermodynamic tables. Note that the two independent properties could be any pair of pressure, temperature, volume, enthalpy, internal energy, and entropy.

To determine properties required in thermodynamic problems, the author has found the engineering equation solver (EES) software [1] as a useful tool. A limited free online version of the software is, at the time of writing

Entropy Analysis in Thermal Engineering Systems
https://doi.org/10.1016/B978-0-12-819168-2.00001-5

this book, also available [2]. EES has been the primary tool for calculating the properties and modeling of most systems discussed in the present book. This will be reiterated in the forthcoming chapters where appropriate.

1.2 Conservation of mass

The principle of mass conservation states that matter is neither created nor destroyed. This principle like many physics laws is empirical; that is, its validity rests on experimental observations. In every process, it is necessary to obey the law of mass conservation. The total amount of matter in a given process is fixed, but it may change from one form to another. For example, consider condensation of steam. In this process, water is initially in gas phase, but then it undergoes a condensation process. At the final state, water is in liquid phase. The conservation of mass requires that the mass of water at its initial state (steam) be equal to the mass of liquid water.

Mass is also conserved in chemical reactions. For example, consider oxidation of hydrogen. The product of reaction is water. The reactants (O_2 and H_2) no longer exist after the reaction and a new product (H_2O) is formed. The conservation of mass requires that the sum of the masses of oxygen and hydrogen be equal to the mass of water. In general, the total mass of reactants should equal the total mass of products in a chemical reaction in accordance with the law of mass conservation.

In mathematical form, the conservation of mass applied to an open system undergoing a steady-state process with n inlet and m outlet ports is written as follows.

$$\sum_{i=1}^{n} \dot{m}_i = \sum_{j=1}^{m} \dot{m}_j \tag{1.1}$$

where $\dot{m}$ denotes mass flow per unit of time, or mass flowrate.

If the system undergoes an unsteady (transient) process, the conservation of mass is expressed as

$$\Delta m_{sys} = \sum_{i=1}^{n} m_i - \sum_{j=1}^{m} m_j \tag{1.2}$$

Eq. (1.2) states that in a transient process taking place over a given time period, t, the net change in the mass of the system, Δm_{sys}, is equal to the sum of the masses entering the system through n inlet port minus the sum of the masses leaving the system through m outlet ports.

1.3 Conservation of energy

The principle of energy conservation states that energy is neither created nor destroyed. It may transform from one type to another. Like the mass conservation principle, the validity of the conservation of energy relies on experimental observations; thus, it is an empirical law. No experiment has violated the principle of energy conservation yet. The common forms of energy include thermal, electrical, chemical, mechanical, kinetic, and potential. It may also be stated that the sum of all kinds of energy is constant.

$$\sum_k E_k = \text{constant} \tag{1.3}$$

where E denotes energy and subscript k refers to the different types of energy.

Many engineering applications involve transformation of energy between two or three types only. For instance, in dynamics problems, the conservation of energy accounts for two types of energy, i.e., kinetic and potential (in some cases frictional work), neglecting the effect of other forms like chemical, thermal, or electrical. In chemical reactions, the conservation of energy includes thermal and chemical energies, and the effect of other forms of energy is ignored. In most thermodynamic problems, the principle of energy conservation applied to nonreactive systems accounts for thermal and mechanical energies.

1.4 First law of thermodynamics

The two main laws of thermodynamics formulated in 19th century were first introduced as the main principles of the Mechanical Theory of Heat. The first law is indeed an expression of the energy conservation principle that accounts for transformation of energy in the form of heat and work. The first law, in the words of Clausius [3], is the principle of the equivalence of heat and work.

The mathematical expression of the first law applied to a closed system with a fixed quantity of mass can be written as follows.

$$\Delta U = Q - W \tag{1.4}$$

where U is the internal energy, Q is the amount of heat transferred to the system, and W denotes the work done by the system on its surrounding.

Eq. (1.4) states that the change in the internal energy of a closed system in a thermodynamic process equals the heat received by the system from an external source minus the work performed by the system. Note that the sign convention used in Eq. (1.4)—and throughout this book—is that the heat transferred to the system is positive and the heat transferred from the system to its surrounding is negative. Further, the work performed by the system on its surrounding is negative, whereas the work done on the system is positive.

The differential form of the first law applied to a closed system is

$$dU = \delta Q - \delta W \tag{1.5}$$

where d is an exact differential, whereas δ denotes an inexact differential.

The first law equation for an open system (control volume) undergoing a steady-state process, which can exchange mass with its surroundings through n inlet and m outlet ports, obeys

$$\dot{Q} - \dot{W} = \sum_{j=1}^{m} \dot{m}_j h_j - \sum_{i=1}^{n} \dot{m}_i h_i + \Delta \dot{E}_{ke} + \Delta \dot{E}_{pe} \tag{1.6}$$

where $\dot{Q}$ denotes the rate of heat transfer, $\dot{W}$ is the power (rate of work), h is the specific enthalpy, $\Delta \dot{E}_{ke}$ denotes the difference between the kinetic energies of the outflows and inflows, and $\Delta \dot{E}_{pe}$ accounts for the difference between the potential energies of the outflows and inflows.

In many thermodynamic applications, the effects of kinetic energy and potential energy are neglected. Eq. (1.6) then reduces to

$$\dot{Q} - \dot{W} = \sum_{j=1}^{m} \dot{m}_j h_j - \sum_{i=1}^{n} \dot{m}_i h_i \tag{1.7}$$

For an open system undergoing a transient process over a finite time, the first law equation reads

$$Q - W = \sum_{j=1}^{m} m_j h_j - \sum_{i=1}^{n} m_i h_i + \Delta U \tag{1.8}$$

In the case of no inlet flow ($m_i = 0$) and no outlet flow ($m_j = 0$), Eq. (1.8) reduces to Eq. (1.4). For an adiabatic process ($Q = 0$) and in the absence of work, the change in the energy of the system depends solely on the outgoing and incoming enthalpy flows.

1.5 Second law of thermodynamics

The second law is based on the observations that thermodynamic processes proceed spontaneously in certain directions. It is an empirical law as no experimental observation has violated its validity yet. A well-known statement of the second law credited to Clausius says that heat cannot be transferred from a cooler body to a warmer body without an external effect. Based on the observations in nature, the spontaneous heat transfer process takes place in one direction only: from a hot region to a cold region. An engineering application of the Clausius statement of the second law is refrigerator. To maintain the inside of a refrigerator cool, heat should be transferred from the interior part of the refrigerator to the surrounding with the use of a compressor (i.e., the external effect).

Another well-known statement of the second law is credited to Lord Kelvin (William Thomson) and Max Plank, which says that it is impossible to construct an engine, which receives heat and converts it all to work. In other words, the efficiency of a heat engine may never reach 100%. Any attempt made in the past by engineers or inventors to violate the Kelvin-Planck statement of the second law was unsuccessful. Indeed, earlier than Clausius, Kelvin, and Plank, Sadi Carnot had understood the second law [4]. As an engineer, Carnot's main goal was to design an engine that would produce a maximum work from a given quantity of heat. His efforts led to the invention of an engine that would operate on a cycle consisting of two isothermal and two adiabatic processes.

The Carnot cycle played an important role in formulation of the second law by Thomson and Clausius, which led to the introduction of entropy by the latter as a new thermodynamic property. Clausius showed that in a reversible process, the change in the entropy of a system, S, is related to an infinitesimal heat transfer δQ by the following relation.

$$dS = \left(\frac{\delta Q}{T}\right)_{rev} \tag{1.9}$$

where T denotes the absolute temperature of the system and has units of Kelvin.

Eq. (1.9) is the analytical formulation of the second law in differential form. It explicitly shows that the change in entropy depends solely on the amount of heat transfer and temperature. To calculate the entropy difference between two states of a system, Eq. (1.9) is integrated along a reversible path.

$$S_2 - S_1 = \int_1^2 \left(\frac{\delta Q}{T}\right)_{rev} \tag{1.10}$$

Either of Eqs. (1.9) and (1.10) is the quantitative definition of entropy [5]. For example, the amount of heat required to evaporate 1 kg of water at 100°C is 2256 kJ. The change in the entropy of 1 kg of saturated liquid water undergoing an evaporation process at 100°C is calculated as follows.

$$S_2 - S_1 = \int_1^2 \left(\frac{\delta Q}{T}\right)_{rev} = \frac{Q}{T} = \frac{2256\ (\text{kJ})}{373.15(\text{K})} = 6.046\ \text{kJ/K}$$

where S_1 is the entropy of liquid water and S_2 denotes the entropy of water vapor at 100°C. The specific entropy of saturated liquid water at 100°C is 1.307 kJ/kgK. So, the entropy of 1 kg of saturated water vapor is determined as

$$S_2 = (1\ \text{kg})(1.307\ \text{kJ/kgK}) + 6.046 = 7.353\ \text{kJ/K}$$

This result can be verified using thermodynamic tables or a property software. In the example, the water temperature was constant throughout the process. Often, the temperature varies along the process and the integral in Eq. (1.10) is evaluated using analytical or numerical integration techniques. For instance, the change in the entropy of 1 kg of water at atmospheric pressure whose temperature increases from 20°C to 60°C is determined as follows.

Use $\delta Q = mc_p dT$ in Eq. (1.10) and perform integration between 293.15 and 333.15 K assuming a specific heat of 4.18 kJ/kgK for water. The change in the entropy of the water is thus obtained as

$$S_2 - S_1 = \int_{T_1}^{T_2} \frac{mc_p dT}{T} = mc_p \ln\frac{T_2}{T_1} = (1)(4.18)\ln\frac{333.15}{293.15} = 0.535\ \text{kJ/K}$$

A subtle but important note from the preceding two examples is that the change in the entropy of a system (e.g., water in the above examples) is determined assuming that the heat transfer takes place *reversibly*. It should also be noted that we used Eq. (1.10) to determine the change in the entropy of water only without regard to the heat source as the immediate surrounding of the system.

In a like manner, one may also calculate the change in the entropy of the surrounding that provided heat to the water—see Section 1.7. Moreover, it can be deduced from Eq. (1.10) that the entropy of system will increase if it receives heat, and the system entropy will decrease if it loses heat [6]. In the latter case, the change in the system entropy will be negative contrary to the former where the entropy change is positive, as demonstrated in the above examples.

1.6 Third law of thermodynamics

Eq. (1.10) enables one to calculate the difference in the entropy of a substance at two different states. To determine the entropy at any state, Eq. (1.10) needs to be integrated from a reference state, which is defined using the third law. Known also as the Nernst theorem [7], the third law says that the entropy of a system at absolute zero temperature is zero [8]. Unlike the first and the second laws, the nature of the third law is not based on experimental observations; it is rather a postulation.

The entropy of a system at any given state, S, where the temperature is T is thus obtained by substituting $S_1 = 0$ at $T = 0$ as the lower limit of the integral in Eq. (1.10).

$$S = \int_0^T \left(\frac{\delta Q}{T} \right)_{rev} \tag{1.11}$$

Notice that the integral in both Eqs. (1.10) and (1.11) is taken along a *reversible* path from any state at zero Kelvin to a state at which the temperature is T.

1.7 Entropy generation

The concept of reversibility introduced by Carnot refers to a process that spontaneously takes place from state A to state B, and from B to A (i.e., in reverse direction) without an external effect. For example, if a quantity of heat is transferred along a reversible path from a warmer body to a cooler body, the same quantity of heat could be transferred from the cooler to the warmer body without a need to an external effect (i.e., power).

Natural processes, as we know, occur in certain directions. To reverse the direction of a process requires an external force that would not be needed in the spontaneous direction. In other words, natural processes are irreversible. The concept of entropy generation is a consequence of the irreversibility in thermodynamic processes. It was first introduced by Clausius as the *uncompensated transformation* who viewed the second law as the principle of the equivalence of transformations.

To alter the entropy of a system would yield a change in the entropy of its immediate surrounding. Let us consider the second example of Section 1.5. The amount of heat required to increase the temperature of water from 20°C to 60°C is $Q = mc_p\Delta T = (1)(4.18)(60-20) = 167.2$ kJ. Suppose this amount of heat is supplied from a condensing steam at 100°C. Given the evaporation enthalpy of 2256 kJ/kg for water at 100°C, about 0.074 kg steam should be condensed to provide 167.2 kJ heat.

The change in the entropy of the condensing steam at the constant temperature of 100°C is calculated as follows.

$$\Delta S_{steam} = S_2 - S_1 = \frac{-167.2}{373.15} = -0.448\,\text{kJ/K}$$

The negative sign indicates that the heat is extracted from the steam. Thus, the change in the entropy of the steam is also negative.

If we now consider the net entropy change (of the system and its surrounding), we find

$$\Delta S_{net} = \Delta S_{water} + \Delta S_{steam} = 0.535 - 0.448 = 0.087\,\text{kJ/K}$$

That is, the process of heating water from 20°C to 60°C where the source of heat is the steam condensing at 100°C leads to a net increase of 0.087 kJ/K in entropy. This net entropy increase is referred to as the entropy generation.

1.7.1 Entropy generation in closed systems

The relation for the entropy generation of a system with a fixed mass can now be presented by generalization of the example that we just discussed above.

$$\Phi = \Delta S_{system} + \Delta S_{surrounding} \tag{1.12}$$

where the change in the entropies of the system and the surrounding can be evaluated using Eq. (1.10). If the system receives heat from its surrounding, the first term on the right-hand side of Eq. (1.12) yields a positive value, whereas the second term leads to a negative value. Conversely, if the system loses heat to its surrounding, the first term yields a negative value and the second term leads to a positive value. In either case, Eq. (1.12) will always have a positive quantity.

Eq. (1.12) may be expressed in alternative forms depending on whether the temperature of the system or surrounding is constant. For example, if the system receives an amount of heat Q from a heat source (the surrounding) maintained at a constant temperature, the entropy generation is determined as

$$\Phi = S_2 - S_1 - \frac{Q}{T_s} \tag{1.13}$$

where $S_2 - S_1$ is the increase in the system entropy and T_s denotes the heat source temperature.

If the system temperature also remains constant during the process, Eq. (1.13) becomes

$$\Phi = \frac{Q}{T_{sys}} - \frac{Q}{T_s} \tag{1.14}$$

Eq. (1.14) is indeed a simple explanation for the generation of entropy. As the heat flows from the surrounding (heat source) to the system, the surrounding temperature should be greater than the system temperature; i.e., $T_{sys} < T_s$. From this, we have $(1/T_{sys} - 1/T_s) > 0$. Also, because $Q > 0$, we conclude that $Q(1/T_{sys} - 1/T_s) > 0$, and thus $\Phi > 0$.

It should be remembered that the surrounding of a system is the region at the vicinity of the boundary of the system, which may have energy interaction (work, heat, or both). If one takes the system and its surrounding as a new system with no external interactions, the new system can be treated as an *isolated* system. Hence,

$$\Phi = \Delta S_{system} > 0 \tag{1.15}$$

Eq. (1.15) states that for an isolated system whose state changes from 1 to 2, the entropy at the final state 2 will be higher than that at the initial state 1. For example, consider a 2-kg block of carbon steel at 90°C that is dropped into a 5-L perfectly insulated container filled with water at 20°C. The system of block + water is an isolated system, which reaches a thermal equilibrium once the temperature of the block and the water becomes the same.

The thermal equilibrium temperature can be determined from the first law equation, i.e., Eq. (1.4). For the system of block + water, we have

$$\Delta U = 0 \rightarrow \Delta U_{block} + \Delta U_{water} = 0 \rightarrow (mc)_{block}\left(T_{eq} - 90\right) + (mc)_{water}\left(T_{eq} - 20\right) = 0$$

Solving the equation with a specific heat of 0.49 kJ/kg K for the block and 4.18 kJ/kg K for the water yields $T_{eq} = 23.14\,^\circ\mathrm{C}$. The entropy generation can now be determined using Eqs. (1.15) and (1.10).

$$\Phi = \Delta S_{block} + \Delta S_{water} = (mc)_{block} \int_{90+273.15}^{T_{eq}} \frac{dT}{T} + (mc)_{water} \int_{20+273.15}^{T_{eq}} \frac{dT}{T}$$
$$= (2)(0.49) \ln \frac{23.14+273.15}{90+273.15} + (5)(4.18) \ln \frac{23.14+273.15}{20+273.15}$$
$$= 0.023 \, \text{kJ/K}$$

1.7.2 Entropy generation in open systems

For an open system with n inlet and m outlet ports that undergoes a steady-state operation, the rate of entropy generation is determined using Eq. (1.16).

$$\dot{\Phi} + \sum_k \frac{\dot{Q}}{T_k} = \sum_{j=1}^{m} \dot{m}_j s_j - \sum_{i=1}^{n} \dot{m}_i s_i \quad (1.16)$$

where the second term on the left-hand side of Eq. (1.16) accounts for the net change in the entropy of the surroundings [9], and $\dot{m}s$ is the entropy rate crossing the system boundary due to the mass flow.

For example, in a nonmixed adiabatic heat exchanger with one hot fluid and one cold fluid, Eq. (1.16) reduces to

$$\begin{aligned} \dot{\Phi} &= (\dot{m}_h s_{h,o} + \dot{m}_c s_{c,o}) - (\dot{m}_h s_{h,i} + \dot{m}_c s_{c,i}) \\ &= \dot{m}_c (s_{c,o} - s_{c,i}) + \dot{m}_h (s_{h,o} - s_{h,i}) \end{aligned} \quad (1.17)$$

where the subscripts h, c, i, o denote hot, cold, inlet, and outlet, respectively.

1.8 Combined first and second laws

The first analytical expression for the combined first and second laws was given by Clausius [3]. It can be obtained by eliminating δQ between Eqs. (1.5) and (1.9). Hence,

$$dU = TdS - \delta W \quad (1.18)$$

For a compressible fluid, if the work done is due to the fluid pressure only, the infinitesimal work can be represented by $\delta W = pdV$ where p represents the pressure and V the volume, and Eq. (1.18) is rewritten as follows.

$$dU = TdS - pdV \quad (1.19)$$

Eq. (1.19) provides a relation between the thermodynamic properties of the system. An alternative expression may be obtained using the differential form of Eq. (1.13) as

$$d\Phi = dS - \frac{\delta Q}{T_s} \tag{1.20}$$

Thus, the combined first and second laws relation becomes

$$dU = T_s dS - pdV - T_s d\Phi \tag{1.21}$$

Note that in Eq. (1.19) T is the system temperature whereas in Eq. (1.21) T_s denotes the surrounding temperature. Also, both Eqs. (1.19) and (1.21) are valid for compressible fluids (gases) that exchange heat with the surroundings and perform work due to the fluid pressure.

For an ideal gas whose state equation is given by

$$pV = nRT \tag{1.22}$$

where n denotes the number of moles and R is the universal gas constant, the internal energy is a function of temperature only; that is, $dU = nc_v dT$. In this case, Eq. (1.19) can be expressed as

$$c_v \frac{dT}{T} = ds - R\frac{dV}{V} \tag{1.23}$$

where c_v denotes the specific heat at constant volume and s is the specific entropy.

References

[1] http://www.fchart.com/ees/.
[2] https://www.irc.wisc.edu/properties/.
[3] R. Clausius, The Mechanical Theory of Heat, Translated by W. R. Brown, MacMillan & Co., London, 1879.
[4] S. Carnot, R.H. Thurston (Ed.), Reflections on the Motive Power of Heat, second ed., Wiley, New York, 1897.
[5] B.F. Dodge, Chemical Engineering Thermodynamics, McGraw-Hill, New York, 1944.
[6] E.F. Obert, Thermodynamics, first ed., McGraw-Hill, New York, 1948.
[7] W. Nernst, Experimental and Theoretical Applications of Thermodynamics to Chemistry, Charles Scribner's Sons, New York, 1907.
[8] E. Fermi, Thermodynamics, Dover Publications Inc., New York, 1956
[9] J.M. Smith, H.C. Van Ness, M.M. Abbott, M.T. Swihart, Introduction to Chemical Engineering Thermodynamics, eighth ed., McGraw-Hill, New York, 2018.

CHAPTER TWO

Birth and evolution of thermodynamics

2.1 Introduction

The history of thermodynamics and the path through which it evolved has rarely been discussed in modern texts. The need for a careful examination of the history emerges for two main reasons. First, the second law whose formulation led to entropy is a challenging subject for many since the present tutorial method of entropy taught in thermodynamics classes is not purely scientific. The author found it necessary to take a journey by reviewing the works of the founders and pioneers of thermodynamics. The best teachers to learn a theory or concept from are those who discovered and developed that theory. Second, there appears to be claims and inaccurate statements (i) underscoring the true contribution of one by unfairly giving credits to else one, and (ii) misrepresenting a theory or concept through personal (mis)interpretation that deviates substantially from the original discoverer of a phenomenon or founder of that theory. The reason for such misrepresentation could be prejudice or due to a lack of sufficient knowledge of the true history of the science of thermodynamics.

The chief objective of the present chapter is to provide the reader an overview of the evolution of the science of thermodynamics during the 19th century, which according to Tait [1], had become an established branch of science by 1877. The author has found the historical journey of thermodynamics fascinating and more importantly educational. It is believed that once the progression of a theory is followed step by step as it occurred, and the concepts originated thereafter are truly understood as enunciated by the original founders, it would be less challenging to comprehend difficult concepts like entropy. Every attempt has been made to carefully examine numerous sources to ensure the accuracy of the contents presented. The events will be discussed chronologically in three sections. It is hoped that the present chapter will enable the reader to have a clearer picture of how the

Entropy Analysis in Thermal Engineering Systems
https://doi.org/10.1016/B978-0-12-819168-2.00002-7

fundamental thermodynamic laws, especially the second law, were discovered and formulated analytically. For this, we will only point out those who substantially contributed to the foundation and development of the *Mechanical Theory of Heat*.

2.2 Before 1800

The science of thermodynamics as we know today was undiscovered during the industrial revolution and throughout the 18th century. Steam engines were in use in England, which were then spread out to other European countries, but their theory was not understood well. The birth of modern thermodynamics is attributed to the work of Sadi Carnot although his investigation was based upon the caloric theory according to which heat was treated as a matter. But what made Carnot's theory plausible was the empirical laws governing the relation between the pressure, volume, and temperature of gases, which were referred to by Carnot as the laws of Mariotte and Gay-Lussac.

2.2.1 Boyle's law

Boyle's law, sometimes also referred to as the law of Boyle-Mariotte, states that the pressure and volume of a given amount of (perfect) gas at constant temperature are inversely related. The law may analytically be expressed as

$$p \propto \frac{1}{V} \tag{2.1}$$

The law given by Eq. (2.1) was first published in 1662 by Robert Boyle who through a series of experiments with air led to conclude that "*as common air when reduced to half its wonted extent, obtained near about twice forcible a spring as it had before; so this thus compressed air being further thrust into half this narrow room, obtained thereby a spring about as strong again as that it last had, and consequently four times as strong as that of the common air*" [2]. Boyle also acknowledged that this law had also been noticed by Richard Townley. Edme Mariotte discovered independently Boyle's law in 1679 [3]. In Carnot's essay, the law given by Eq. (2.1) has been referred to as Mariotte' law.

2.2.2 Charles' law

Charles' law states that the volume of a given quantity of (perfect) gas is proportional to its absolute temperature at constant pressure. In analytical form, the law may be expressed as

$$V \propto T \tag{2.2}$$

This law is credited to Jacques Charles as its discoverer in 1787, although his work was not published. Later, Dalton and Gay-Lussac conducted experiments with air and other gases, which confirmed the law given by Eq. (2.2).

2.2.3 Experiments of Rumford and Davy

The notion of materiality of heat and thus the caloric theory was invalidated by Count Rumford in 1798 and Humphry Davy in the following year. Tait regarded Rumford and Davy as the true discoverer of the first law of thermodynamics whose contribution was unnoticed until Joule re-established the law through a series of comprehensive experiments [4].

As the superintendent of the cannon boring in the workshops of military arsenal at Munich, Rumford noticed a considerable amount of heat gained by a brass gun that was bored in a short time. In his communication to the Royal Society, he presented the details of four experiments, which led him to question:

> *What is heat? Is there any such thing as an igneous fluid? Is there anything that can with propriety be called caloric? We have seen that a very considerable quantity of heat may be excited in the friction of two metallic surfaces and given off in a constant stream or flux, in all directions, without interruption or intermission, and without any signs of diminution, or exhaustion ... it appears to me to be extremely difficult, if not quite impossible, to form any distinct idea of anything, capable of being excited, and communicated, in the manner the heat was excited and communicated in these experiments, except it be motion [5].*

The experiment of Davy consisted of two pieces of $6 \times 2 \times 2/3$ in. ice, which were rubbed together and then almost entirely melted. He wrote: "*The fusion took place only at the plane of contact of the two pieces of ice, and no bodies were in friction but ice. From this experiment it is evident that ice by friction is converted into water, and according to the supposition [of materiality of heat] its capacity is diminished; but it is a well-known fact, that the capacity of water for heat is much greater than that of ice; and ice must have an absolute quantity of heat added to it, before it can be converted into water. Friction consequently does not diminish the capacities of bodies for heat.*" From this argument, Davy concluded the immateriality of heat. He then elucidated that "*Heat ... may be defined a peculiar motion, probably a vibration, of the corpuscles of bodies ... It may with propriety be called the repulsive motion*" [6].

Tait, in appraising the works of Rumford and Davy, writes: "*Notice how distinctly these two great leaders were men who based their work directly upon experiment. There is no a priori guessing, or anything of that kind, about either Rumford's or Davy's work. They simply set to work to find out what heat is. They did not speculate on what it might be*" [4].

2.3 Between 1800 and 1849

In the commencement of the 19th century, John Dalton noted the rise of air temperature when it was compressed, and a drop of temperature when the air was expanded. He communicated a series of 11 experiments to the Society of Manchester on June 27, 1800 [7]. A year later, Dalton read four papers before the same society. In the first paper, Dalton furnished the law of mixed gases on the basis of the supposition that "*The particles of one elastic fluid may possess no repulsive or attractive power, or be perfectly inelastic with regard to the particles of another: and consequently the mutual action of the fluids be subject to the laws of inelastic bodies*"; Ref. [8], p. 543.

According to Dalton's law, the pressure of a gaseous mixture comprising k nonreactive gases is equal to the sum of the partial pressures of the constituent gases.

$$p_{\text{mix}} = p_1 + p_2 + \ldots + p_k \tag{2.3}$$

In the fourth paper, Dalton investigated the expansion of gases with the rise of temperature. The subject was previously studied by not only Charles, as mentioned in the preceding section, but also other French investigators like Berthollet, Guyton de Morveau, and Duvernois, as stated by Dalton. Through a series of experiments with air and some other gases, he was led to ascertain that "*all elastic fluids expand equally by heat—and that for any given expansion of mercury, the corresponding expansion of air is proportionally something less, the higher the temperature*"; Ref. [8], p. 600.

In a paper published in 1802, Joseph Gay-Lussac presented the results of his research on expansion of gases and vapors along with an ample historical sketch summarizing the efforts of earlier investigators of the subject, including Amontons, Nuguet, Lahire, Stancari, Deluc, Roy, Saussure, Priestley, Monge, Berthollet, Vandermonde, Guyton de Morveau, and Duvernois [9]. Gay-Lussac found great discrepancies in the results of the previous experimenters. He also acknowledged the (unpublished) findings of Charles: "*Although I had noticed a great many times that the gases oxygen, nitrogen, hydrogen, carbonic acid and atmospheric air expand to the same extent from 0° to 80°, citizen Charles had, fifteen years before, discovered the same property in those gases; but, never having published his results*" [10].

Upon concluding a series of experiments with several gases such as air, oxygen, hydrogen, nitrogen, Gay-Lussac summarized his findings as follows: "*All gases, whatever their density or quantity of water which they hold in solution, and all vapors expand to the same extent for the same degree of heat. In the case of the permanent gases, the increase of volume which each of them suffers between the degree of melting ice and that of boiling water amounts to 80/213.33 of the original volume, for a thermometer divided in 80 parts, or 100/266.66 of the same volume, for a centigrade thermometer.*" The coefficient of expansion calculated from the Gay-Lussac work is 0.00375, which was later corrected to 0.003665 based on more accurate experiments of Magnus and Regnault.

2.3.1 Carnot's contribution

It was over a century that steam engines had already been in use in mines, ships, and many other applications, but little was known about the theory of heat engines. To the best knowledge of the author, the earliest commercial steam engine was patented by Thomas Savery in 1698, which was employed for pumping water [11]. Further improvements in the design of steam engines were introduced by Thomas Newcomen and James Watt.

It was Sadi Carnot who initiated investigating the theory of heat engines in the early 19th century, which led to a 118-page manuscript published in 1824 [12]. In the introductory portion of his book, when acknowledging the earlier inventors, he wrote: "*If the honor of a discovery belongs to the nation in which it has acquired its growth and all its developments, this honor cannot be here refused to England*" (see p. 6 in Ref. [12]. The English translation is taken from Ref. [13], p. 41). The fundamental question, which laid down Carnot's investigation, was whether there is a limit for the power generated in a heat engine. Carnot developed his arguments based on the caloric theory, although it was already invalidated by Rumford and Davy as discussed in Section 2.2.3. He then established the following supposition:

> *The production of motive power is then due in steam-engines not to an actual consumption of caloric, but to its transportation from a warm body to a cold body, that is, to its re-establishment of equilibrium ... the production of heat alone is not sufficient to give birth to the impelling power: it is necessary that there should also be cold; without it, the heat would be useless; [13], p. 46.*

It can be inferred from these statements that Carnot was somewhat aware of the second law. The first axiom that he used in demonstration of the design of perfect engine is that the production of power would be possible wherever there exists a temperature difference. Carnot had also understood that motion caused by heat can only be due to a change of volume. In describing the characteristics of perfect engine that would produce a maximum power, he wrote ([13], p. 57):

> *The necessary condition of the maximum is, then, that in the bodies employed to realize the motive power of heat there should not occur any change of temperature which may not be due a change of volume. Reciprocally, every time that this condition is fulfilled the maximum will be attained.*

Carnot viewed establishment of a thermal equilibrium without production of power as an actual loss. He suggested that every temperature difference that is not due to a volume change or chemical reaction like combustion should be avoided to maintain the condition of maximum power. Contrary to the commonly accepted notion among the present scholars according to which combustion process is viewed as a large source of "loss" in a power producing device, Carnot had assumed the absence of any chemical reaction in his demonstration (Ref. [13], p. 56):

> *The chemical action which takes place in the furnace is, in some sort, a preliminary action, -an operation destined not to produce immediately motive power, but to destroy the equilibrium of the caloric, to produce a difference of temperature which may finally give rise to motion.*

Carnot then devised the operation of an engine that would operate in a cycle while communicating with two thermal reservoirs of fixed but different temperatures. Any change in the temperature of the working substance may occur due to compression or expansion only, and the heat exchange processes take place at the reservoirs' temperatures. To prove that his proposed engine produces a maximum power, Carnot employs a rather philosophical reasoning. He first argues that if the operation of the engine is reversed; that is, to operate it as a refrigerator, the mechanical and thermal effects of the engine would be identical to those if it were operated backward, i.e., as a refrigerator. He then reasons that if there were any method (i.e., engine) to produce a greater power than that of his proposed engine, it would be enough to spend a portion of this power to transfer heat from the cold reservoir to the hot reservoir by executing the Carnot engine backward, whereby a combination of the two engines would lead to a perpetual machine.

The above method of demonstration, despite founded erroneously on caloric theory, has since been used by not only his successors like Clausius, but in almost all introductory thermodynamics textbooks published untill the present day. In Chapter 3, we will treat this subject in detail and discuss the shortcomings of the rational reasoning introduced by Carnot. Nevertheless, what is remarkable with respect to the Carnot's contribution unappreciated, perhaps, in all textbooks is the analytical formulation he presented using the laws of Boyle-Mariotte and Dalton-Gay Lussac; see pp. 74–76 in Ref. [12] with the English translation given in Appendix B of Ref. [13]. Carnot expressed the combined laws, i.e., Eqs. (2.1) and (2.2), in a single equation as

$$V = c\frac{t+267}{p} \tag{2.4}$$

where V is the volume, p the pressure, t the temperature (in centigrade), c a constant, and the number 267 is the inverse of the expansion coefficient obtained from the experiments of Gay-Lussac.

Eq. (2.4) is an early version of the ideal gas equation which was used by Carnot to establish a relation for the isothermal expansion work of air from a unit of volume to any given volume like V at constant temperature t.

$$r = c(267 + t)\ \ln V \tag{2.5}$$

where r, in Carnot's notations, refers to the quantity of work produced in the process.

Carnot then derived another relation for the quantity of heat, designated by e, required to maintain the temperature constant during the expansion process.

$$e = \frac{c}{F'}\ \ln V \tag{2.6}$$

where F' is a function that depends on temperature only.

A careful examination of Carnot's analysis reveals that he had a correct understanding of the principle of the equivalence of heat and work. It is from this analytical formulation that he was led to state some of his propositions; for instance "When a gas varies in volume without change of temperature, the quantities of heat absorbed or liberated by this gas are in arithmetical progression, if the increments or the decrements of volume are found to be in geometrical progression" (Ref. [13], p. 81), or "The motive power of heat is independent of the agents employed to realize it; its quantity is fixed solely by the temperatures of the bodies between which is effected, finally, the transfer of the caloric" (Ref. [13], p. 68). It is interesting to note that Carnot had carefully shown the limitation and the range of the validity of his theory. He wrote:

> *The theories that we deduce here would not perhaps be exact if applied outside of certain limits either of density or temperature. They should be regarded as true only within the limits in which the laws of Mariotte and of Gay-Lussac and Dalton are themselves proven (Ref. [13], p. 80).*

2.3.2 Clapeyron's work

Carnot's work was remained unnoticed for a decade after its publication. It was then Emile Clapeyron who gave a recognition to the Carnot's contribution and presented graphically the operation of the Carnot cycle on a pressure-volume diagram [14]. The analysis of Clapeyron also employs the ideal gas law, which he expressed as

$$pV = R(t+267) \tag{2.7}$$

where $R = \frac{p_0 V_0}{t_0 + 267}$ and the subscript "0" denotes a reference state.

By applying differential calculus, Clapeyron formulated the mechanical work and the heat required to produce this effect in a Carnot engine operating with a perfect gas. He showed that the ratio of the work produced to the heat supplied obeys $\frac{dt}{C}$ where C, a function of temperature, is the inverse of function F' in Carnot's notations; see Eq. (2.6). We will later show that C is indeed the absolute temperature, T. Clapeyron also showed that the difference between the specific heat at constant pressure and that at constant volume of (perfect) gases is $\frac{R}{267+t}C$; see pp. 170–171 in Ref. [14].

2.3.3 Poisson's equations

In 1833, Siméon Denis Poisson presented equations relating the pressure, volume, and temperature of a perfect gas undergoing an adiabatic process at two different states [15].

$$\frac{p}{p'} = \left(\frac{\rho}{\rho'}\right)^{\gamma} \tag{2.8}$$

$$\frac{t + 266.67}{t' + 266.67} = \left(\frac{\rho}{\rho'}\right)^{\gamma - 1} \tag{2.9}$$

where ρ denotes the density and γ is the ratio of the specific heats.

Poisson derived the two equations by assuming γ as a constant parameter, and by using the combined gas laws of Boyle-Dalton or Mariotte-Gay Lussac, which he expressed as

$$p = k\rho(1 + \alpha t) \tag{2.10}$$

He noted that α and k are independent of density and temperature, where $\alpha = 0.00376$ being the same for all gases, but k to be given for a particular gas.

2.3.4 Experiments of Joule

By the end of the first half of the 19th century, the immateriality of heat had already faded away and the principle of the equivalence of heat and work as a fundamental thermodynamic law had been accepted by the scientific community. Numerous investigators contributed to the realization of and proving the truth of this law of nature. After the pioneering experiments of Rumford and Davy (see Section 2.2.3), it was enunciated by Davy in 1812 that "The immediate cause of the phenomenon of heat, then, is motion, and the laws of its communication are precisely the same as the laws of the communication of motion" [16]. The merit of discovery and foundation of the first law should also be credited to several other ingenious men like Marc Seguin, Julius Robert von Mayer, James Prescott Joule, and Ludwig Colding who had independently realized and expressed the truth of the equivalence of work and heat.

It is however Joule who deserves an extra credit and recognition. He conducted numberless experiments with a great caution that provided a solid foundation to support the validity of the equivalence of heat and work as a general natural law. For this, Tyndall has recognized Joule as the "*experimental demonstrator of the equivalence of heat and work*" [17]. In describing Joule's work, Youmans writes: "*Mr. Joule's experiments are of extreme delicacy—he tabulates to the thousandth part of a degree of Fahrenheit, and a large number of his thermometric data are comprehended within the limits of a single degree. Other experimenters have given very different numerical results, but the general opinion seems to be that the numbers given by Mr. Joule are the nearest approximation to the truth yet obtained*" [18].

Before enunciating his early findings on the mechanical equivalent of heat in 1843, Joule had conducted experiments on electromagnetic forces and voltaic electricity, which led him to establish the law of heat production by electricity. Subsequently, an abstract of his work was published in the Proceedings of the Royal Society in December 1840 [19]. He designed apparatuses to experiment the quantity of heat evolved during the electrolysis of water, combustion of metals, by an iron bar rotating under magnetic effect, agitation of liquids, and friction of fluids. Some of the experimental efforts of Joule on the generation of heat by the friction of fluid will be described in Chapter 4.

After years of labor, in a letter communicated to the editors of *Philosophical Magazine* in 1862, Joule wrote "I do not wish to claim any monopoly of merit. Even if Rumford, Mayer, and Seguin had not produced their works, justice would still compel me to share with Thomson, Rankine, Helmholtz, Holtzman, Clausius, and others, whose labours have not only given developments and applications of the dynamical theory which entitle them to merit as well as their predecessors in these inquiries, but who have contributed most essentially in supporting it by new proofs" [20]. It is evident that not only was Joule a highly skilled scientist and the experimental demonstrator of the first law, but also a humble man who freely admired the effort and contribution of other fellow philosophers, at home and foreign countries, to the development of the science of thermodynamics.

2.3.5 Absolute temperature scale

In a paper published in 1848 [20], William Thomson (Lord Kelvin) begins with a statement that the theory of thermometry is far from a satisfactory state and urges the need for a principle to serve as a foundation for absolute temperature scale. At the time, air thermometer was employed as a standard scale for measuring temperature. Despite the sufficient accuracy of the thermometers constructed with air or other gases, he argued that "Although we have thus a strict principle for constructing a definite system for the estimation of temperature, yet as reference is essentially made to a specific body as the standard thermometric substance, we cannot consider that we have arrived at an absolute scale."

Thomson then proposes Carnot's theory as a foundation for the absolute temperature scale as the ratio of work-to-heat in Carnot cycle depends on temperature, which was shown by Carnot and later by Clapeyron; see Sections 2.3.1 and 2.3.2. In describing the new scale, Thomson wrote: "*The characteristic property of the scale which I now propose is, that all degrees have the same value; that is, that a unit of heat descending from a body A at the temperature T° of this scale, to a body B at the temperature $(T\text{-}l)^{\circ}$, would give out the same mechanical effect, whatever be the number T. This may justly be termed an absolute scale, since its characteristic is quite independent of the physical properties of any specific substance.*"

However, the above proposed scale did not prevail. Using the experiments of Regnault on the pressure and latent heat of saturated steam at temperatures between 0°C and 230°C, Thomson found large discrepancy between the values of temperature determined based on this hypothesis and that of an air thermometer. In 1854, Thomson proposed a modified definition for the absolute temperature scale: "*the numerical measure of temperature shall be not founded on the expansion of air at a particular pressure, but shall be simply the mechanical equivalent of the thermal unit divided by Carnot's function … the formula for the action of a perfect thermo-dynamic engine expresses that the heat used is to the heat rejected in the proportion of the temperature of the source to the temperature of the refrigerator*" [21]. According to this latter definition, the absolute temperature in units of Kelvin is equal to temperature measured in Centigrade plus a numerical constant.

2.4 Theoretical developments

The theoretical development and formulation of the fundamental laws of thermodynamics is primarily due to Rankine, Thomson, and Clausius. In 1849, a year after proposing a foundation for the absolute temperature scale, Thomson extended Clapeyron's investigation and arrived at the following relation for the amount of work required for production of a unit of heat evolved during the compression of air [22].

$$\frac{W}{Q} = \frac{\mu(1+Et)}{E} \tag{2.11}$$

where W is the amount of work spent to compress the air, Q the heat evolved by the compression, μ Carnot's function, t the temperature, and E the expansion coefficient of air.

Like Carnot and Clapeyron, the theoretical analysis of Thomson was based on the combined gas laws, which he expressed in the following form.

$$pV = p_0 V_0 (1+Et) \tag{2.12}$$

The functional form of μ adopted by Thomson [23] is

$$\mu = J\frac{E}{1+Et} \tag{2.13}$$

where J denotes the mechanical equivalent of a thermal unit.

In explanation of Eq. (2.13), Thomson writes "*It was suggested to me by Mr Joule, in a letter dated December 9, 1848, that the true value of μ might be inversely at the temperatures from zero … This formula* [Eq. (2.13)] *is also adopted by*

Clausius, who uses it fundamentally in his mathematical investigations" [23]. Indeed, Rudolf Clausius derived an equation in 1850 similar to that given by Eq. (2.13) in the following form [24]:

$$C = A(a + t) \tag{2.14}$$

where C denotes the inverse of Carnot's function, A designates the equivalent of heat for the unit of work, i.e., inverse of J in Eq. (2.13), and $a = 273$ is the inverse of the expansion coefficient of air, i.e., 0.003665 determined from the experiments of Magnus and Regnault. The relation between the absolute temperature scale and centigrade, $[K] = [°C] + a$, was given by William Rankine in a paper read on February 4, 1850 before the Royale Society of Edinburgh [25]. Clausius also arrived at the same relation [26] with a difference that in the Clausius relation $a = 273$, whereas in that of Rankine $a = 274.6$.

Rankine, in a subsequent paper, presented an analytical formulation to derive an expression for the efficiency of the Carnot cycle [27].

$$\frac{W}{Q_1} = \frac{T_1 - T_0}{T_1} \tag{2.15}$$

where Q_1 is the total heat supplied, T_1 and T_0 denote the temperature at which heat is supplied to and rejected by the working substance, respectively.

A remarkable conclusion enunciated by Rankine is that "*Carnot's Law is not an independent principle in the theory of heat; but is deducible, as a consequence, from the equations of the mutual conversion of heat and expansive power.*" An equation analogous to Eq. (2.15) was also obtained by Thomson using the combined gas laws; i.e., Eq. (2.12), and Poisson's equations, which was published as a note to a paper authored by Joule on air engines [28].

After presenting a modified definition for the absolute temperature scale, Thomson presented mathematical expressions of the first and second laws of thermodynamics for a system undergoing a reversible process [21].

$$W + J(H_t + H_{t'} + \ldots + H_{t^{n-1}} + H_{t^n}) = 0 \tag{2.16}$$

$$\frac{H_t}{t} + \frac{H_{t'}}{t'} + \ldots + \frac{H_{t^{n-1}}}{t^{n-1}} + \frac{H_{t^n}}{t^n} = 0 \tag{2.17}$$

where W denotes the work done, H_t, $H_{t'}$, …, $H_{t^{n-1}}$, H_{t^n} designate the quantities of heat taken in at temperatures t, t', …, t^{n-1}, t^n, respectively.

Thomson's paper [21] was published in May 1854. In December of that year, Clausius published his fourth memoir [26] in Poggendoff's Annalen in that the first law is expressed as

$$Q = U + A \cdot W \tag{2.18}$$

where, in accordance with the Clausius' notations, Q denotes the heat imparted to the system, U is the sum of the thermal energy and interior work, and W designates the exterior work. Clausius explains that the interior work is due to the forces "which the atoms of a body exert upon each other, and which depend, of course, upon the nature of the body." The exterior work refers to the work done by the system in overcoming the external forces.

In the same paper, Clausius enunciated the *Theorem of the Equivalence of Transformations* and derived the following relation for a system undergoing a cyclical process.

$$N = \int \frac{dQ}{T} \geq 0 \tag{2.19}$$

where N, the *uncompensated transformation* as denoted by Clausius, is positive for all real processes, and zero if the process is reversible. Clausius obtained inequality (2.19) taking the heat given off by the system as positive and the heat imparted to the system as negative. However, in his ninth memoir [29], Clausius adopted an opposite sign convention and therefore expressed the inequality (2.19) as

$$\int \frac{dQ}{T} \leq 0 \tag{2.20}$$

Later in the same memoir, he assigned a new parameter for the term under the integration sign.

$$dS = \frac{dQ}{T} \tag{2.21}$$

He wrote: "*We might call S the transformational content of the body, just as we termed the magnitude U its thermal and ergonal content. But as I hold it to be better to borrow terms for important magnitudes from the ancient languages, so that they may be adopted unchanged in all modern languages, I propose to call the magnitude S the entropy of the body.*"

Integrating Eq. (2.21) over a reversible path from a reference state to an arbitrary state yields

$$S = S_0 + \int \frac{dQ}{T} \tag{2.22}$$

As noted by Clausius, Eq. (2.22) is used for determination of the magnitude of S. For the case of perfect gases, he derived the following differential equations for a unit weight of the substance [30].

$$dU = c_v dT \tag{2.23}$$

$$dS = c_v \frac{dT}{T} + A \cdot R_g \frac{dV}{V} \tag{2.24}$$

where R_g denotes the gas constant. Eq. (2.24) may be obtained by combination of Eqs. (2.21), (2.23) and the differential form of Eq. (2.18), i.e.,

$$dQ = dU + A \cdot dW \tag{2.25}$$

where $dW = pdV$.

Recall the coefficient A in Clausius' formulation that denotes the equivalent of heat for a unit of work. If the heat, work, and energy terms in Eq. (2.25) are described with the same unit (e.g., J), the coefficient A will then vanish. In a book that Clausius published years after his ninth memoir, the coefficient A is dropped off from his formulation. For instance, Eq. (2.24) after integration is expressed as [30]

$$S = S_0 + c_v \ln \frac{T}{T_0} + R_g \ln \frac{V}{V_0} \tag{2.26}$$

where the subscript "0" designates an initial or reference state.

2.5 Remarks

The first and foremost conclusion that can be drawn from the forgoing discussions is that the foundation of thermodynamics is due to the labor of many ingenious men, although the individual contribution may differ from one to another. It would be unfair to entitle mere one or two individuals the merit of discovering or founding the main laws of thermodynamics, which evolved through several decades. The role of ideal gas law in the theoretical development of thermodynamics, overlooked in many sources, should especially be highlighted. The way that the second law is taught in introductory classes does not follow the same logic as it was developed by the founders. This issue will further be discussed in the next chapter.

An example is the absolute temperature scale that is usually introduced without a clear justification. If the ratio of the heat received-to-heat rejected in a Carnot cycle is proportional to the ratio of the corresponding (absolute) temperatures, it is not merely due to adopting a new temperature scale, as the relation between the absolute and old temperature scales had already been discovered. The combined gas laws enabled Carnot, Clapeyron, Thomson, and Clausius to deduce that the ratio of work-to-heat in a reversible engine would be a function of temperature only.

Finally, it is important to note that the legitimacy of a natural law strictly depends upon repeatable and accurate experimental observations. One may have a hypothesis about a phenomenon, but it cannot be justified as a valid theory unless its truth is tested through a series of meaningful experiments. Having said that, perhaps the beautiful science of thermodynamics would not have come to an existence without the experiments of Boyle, Mariotte, Dalton, Gay-Lussac, and Joule.

References

[1] P.G. Tait, Sketch of Thermodynamics, second ed., Edmunston and Douglas, Edinburgh, 1877.
[2] R. Boyle, A Defence of the Doctrine Touching the Spring and Weight of the Air, Chapter VThomas Robinson, London, 1662, 62.
[3] E. Mariotte, Oeuvres de Mr. Mariotte, de l'Académie royale des sciences, Second Essai: De La Nature De L'Air, vol. I, Pierre Vander Aa, Leide, 1717, pp. 148–153.
[4] P.G. Tait, Lectures on Some Recent Advances in Physical Science, second ed., Macmillan & Co, 1876.
[5] C. Rumford, An inquiry concerning the source of the heat which is excited by friction, Philos. Trans. R. Soc. Lond. 88 (1798) 80–102.
[6] T. Beddoes, Contributions of Physical and Medical Knowledge, Principally From the West of England, Printed by Biggs & Cottle for T.N. Longman and O. Rees, Paternoster Row, London, 1799, pp. 16–22.
[7] J. Dalton, Experiments and observations on the heat and cold produced by the mechanical condensation and rarefaction of air, in: Memoirs of the Literary and Philosophical Society of Manchester, vol. V, Cadell & Davies, London, 1802, , pp. 515–526 Part II.
[8] J. Dalton, On the constitution of mixed gases; on the force of steam or vapor from water and other liquids in different temperatures, both in a Torricellian vacuum and in air; on evaporation; and on the expansion of gases by heat, in: Memoirs of the Literary and Philosophical Society of Manchester, vol. V, Cadell & Davies, London, 1802, , pp. 535–602 Part II.
[9] J. Gay-Lussac, Recherches sur la dilatation des gaz et des vapeurs, Ann. Chim. 43 (1802) 137–175.
[10] W.W. Randall, The Expansion of Gases by Heat, American Book Company, Dobbs Ferry, NY, 1902, p. 37.
[11] T. Savery, The Miners' Friend, or an Engine to Raise Water by Fire, Crouch, London, 1702.
[12] S. Carnot, Réflexions sur la Puissance Motrice du Feu et Sur Les Machines Propres à Développer Cette Puissance, Chez Bachelier, Paris, 1824.
[13] S. Carnot, R.H. Thurston (Ed.), Reflections on the Motive Power of Heat, second ed., Wiley, New York, 1897.
[14] E. Clapeyron, Mémoire sur la puissance motrice de la chaleur, Journal de L'Ecole Royale Polytechnique: De l'Imprimerie Royale 14 (1834) 153–190.

[15] S.D. Poisson, Traité de Mécanique, second ed., vol. 2, Bachelier, Paris, 1833, pp. 646–647 Chapter 6.
[16] S.H. Davy, Elements of Chemical Philosophy, Part I, vol. I, Johnson & Co., London, 1812, pp. 94–95
[17] J. Tyndall, On Mayer and the mechanical theory of heat, Phil. Mag. J. Sci. XXIV (1862) 174.
[18] E.L. Youmans, The Correlations and Conservation of Forces, Appleton and Co., New York, 1886, p. 33.
[19] J.P. Joule, The Scientific Papers of James Prescott Joule, Physical Society of London, 1884.
[20] W. Thomson, On an absolute thermometric scale founded on Carnot's theory of the motive power of heat, and calculations from Regnault's observations, Philos. Mag. (1848) 100–106.
[21] W. Thomson, On the dynamical theory of heat, thermo-electric currents: fundamental principles of general thermo-dynamics recapitulated, Trans. Royal Soc. Edinb. 21 (1854) 123–127.
[22] W. Thomson, On account of Carnot's theory of the motive power of heat with numerical results deduced from Regnault's experiment on steam, Trans. Royal Soc. Edinb. 16 (1849) 113–155.
[23] W. Thomson, On the dynamical theory of heat with numerical results deduced from Mr Joule's equivalent of a thermal unit and M. Regnault's observations on steam, Trans. Royal Soc. Edinb. 20 (1851) 261–288.
[24] R. Clausius, On the moving force of heat and the laws regarding the nature of heat itself which are deducible therefrom, Phil. Mag. J. Sci. II (1851) 102–119.
[25] W.J.M. Rankine, On the mechanical action of heat especially in gases and vapours, Trans. Royal Soc. Edinb. 20 (1850) 147–190.
[26] R. Clausius, Ueber eine veränderte form des zweiten hauptsatzes der mechanischen wärmetheoriein, Ann. Phys. Chem. 93 (1854) 481–506.
[27] W.J.M. Rankine, On the economy of heat in expansive machines, Trans. Royal Soc. Edinb. 20 (1851) 205–210.
[28] J.P. Joule, On the air engine, Philos. Trans. Royal Soc. Lond. 142 (Part II) (1852) 65–82.
[29] R. Clausius, Ueber verschiedene für die Anwendung bequeme Formen der Hauptgleichungen der mechanischen Wärmetheorie, Ann. Phys. (1865) Bd. cxxv, 353.
[30] R. Clausius, The Mechanical Theory of Heat, Translated by W. R. BrownMacMillan & Co., London, 1879.

CHAPTER THREE

Teaching entropy

3.1 Introduction

The statement of the second law is clear and straightforward: a heat engine cannot convert the entire heat input to work, or heat cannot be transferred from a cooler body to a warmer one without external expenditure. However, the formulation of the second law leads to entropy, which is one of the most challenging concepts to understand by many. We will first review the current common method of teaching the second law and entropy that can be found in most standard textbooks. A part of this common method relies on the Carnot corollaries and their respected proof. It will be shown that the proof of the corollaries inherited from Carnot himself [1] is unsatisfactory and certain objections can be made.

3.2 The common tutorial method

The way entropy is taught in many textbooks includes introduction of several steps: (i) reversible process, (ii) Carnot cycle and principles, (iii) thermodynamic temperature scale, (iv) efficiency of Carnot cycle, (v) Clausius inequality. Then, entropy is introduced as a thermodynamic property along with the principle of entropy increase. This tutorial method has become a standard approach of teaching entropy to undergraduate students for a very long time.

Compared to other properties such as enthalpy, specific volume, internal energy, specific heat, or formation enthalpy, entropy is the only property whose introduction takes a relatively long journey. A whole chapter is usually devoted to present steps (i) through (iv) whereas the inequality of Clausius, entropy, and the principle of entropy increase are discussed in a following chapter. It is not a surprise why grasping entropy is challenging; one needs to first go through several steps to finally arrive at the definition of entropy. On the other hand, steps (ii)–(iv) suffer from certain shortcomings: the proof of Carnot principles and the method of derivation of the Carnot

Entropy Analysis in Thermal Engineering Systems
https://doi.org/10.1016/B978-0-12-819168-2.00003-9

efficiency are not entirely convincing. The Carnot principles—also called Carnot corollaries or propositions—state that

I. the efficiency of an irreversible engine cannot exceed that of the reversible one operating between the same thermal reservoirs,

II. all reversible engines operating between the same two thermal reservoirs possess the same efficiency.

3.2.1 Proof of corollaries

We now briefly review the method of demonstration of Carnot's first corollary, which can be found in nearly all engineering thermodynamics texts. The method was originally presented by Carnot himself to show that his ideal engine is the most efficient among all engines operating between two fixed temperature thermal reservoirs.

Consider an arbitrary engine A (other than Carnot engine) that operates between two thermal reservoirs maintained at fixed temperatures; see Fig. 3.1. The amount of heat transferred from the high-temperature reservoir to engine A is denoted by Q_{HA}, and the amount of heat rejected by the engine to the low-temperature reservoir is Q_{LA}. The first law requires the work production of engine A to be $W_A = Q_{HA} - Q_{LA}$. Next, a Carnot engine C is placed between the same thermal reservoirs. It receives heat Q_{HC} from the high-temperature reservoir, converts a portion of the heat to work W_C, and rejects the unused portion of the heat, Q_{LC}, to the low-temperature reservoir.

Assume that engine A is more efficient than engine C and that the amount of heat transfer from the high-temperature reservoir to both engines is the same, i.e., $Q_{HA} = Q_{HC}$. Thus, $W_A > W_C$ and $Q_{LC} > Q_{LA}$. The

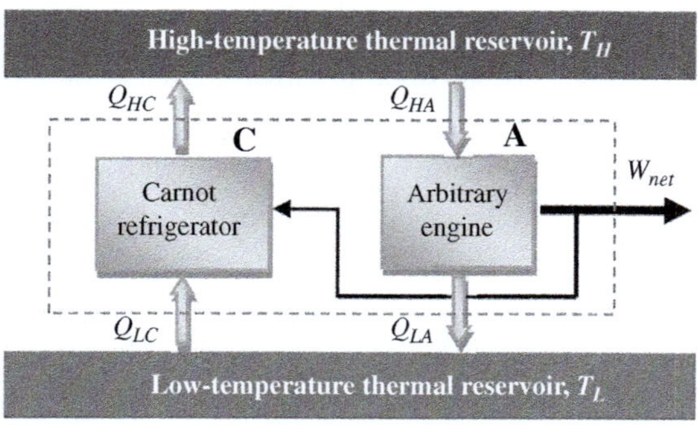

Fig. 3.1 Illustration of an arbitrary engine and a Carnot refrigerator operating between the same thermal reservoirs.

quantities of Q_{HC}, W_C, and Q_{LC} will remain the same if the Carnot engine is operated in reverse direction, i.e., as a refrigerator. Now, consider a combined system of engine A and refrigerator C designated with the dashed rectangle in Fig. 3.1. Since $W_A > W_C$, the work of the refrigerator C can be provided by engine A. The combined system of engine A and refrigerator C would produce a network of $W_{net} = W_A - W_C$. As the amount of heat exchange between the high-temperature reservoir and the engines is assumed to be the same, the net quantity of heat transfer to the combined system from the low-temperature thermal reservoir is $Q_{net,L} = Q_{CL} - Q_{AL}$. The first law requires that $W_{net} = Q_{net,L}$.

The combined system of A + C would violate the second law because it would convert the entire heat received from the low-temperature reservoir to work. Thus, the initial assumption that engine A could produce more work than Carnot engine is incorrect. This is the conclusion of the proof of the first corollary. Thermodynamics textbooks that employ this method of reasoning as a proof of the first corollary also note that Carnot's second corollary is provable in a similar manner. A subtle assumption employed in the argument is that both engine A and the Carnot engine C receive an identical quantity of heat from the high-temperature reservoir. Without the assumption of $Q_{HA} = Q_{HC}$, the argument cannot be used to justify the Carnot's first corollary. It could be presented with a slight modification as follows:

> The efficiency of an irreversible engine cannot exceed that of the reversible one operating between the same thermal reservoirs provided both engines receive an *identical quantity of heat* from the high-temperature reservoir.

3.2.2 Shortcomings of the proof

In this section, we show what objections can be made to the proof of Carnot corollaries discussed in Section 3.2.1. Consider two Carnot engines that operate between the same thermal reservoirs. Assume that both engines receive the same amount of heat $Q_{HC1} = Q_{HC2}$ and that engine C_1 produces a greater work than engine C_2, i.e. $W_{C1} > W_{C2}$. If we follow the same reasoning presented in Section 3.2.1 as the proof of the first corollary, a combined system of engine C_1 and refrigerator C_2 would lead to a violation of the second law.

It is natural to be curious about a situation opposite to that shown in Fig. 3.1. Since both C_1 and C_2 are reversible engines, C_1 is executed in

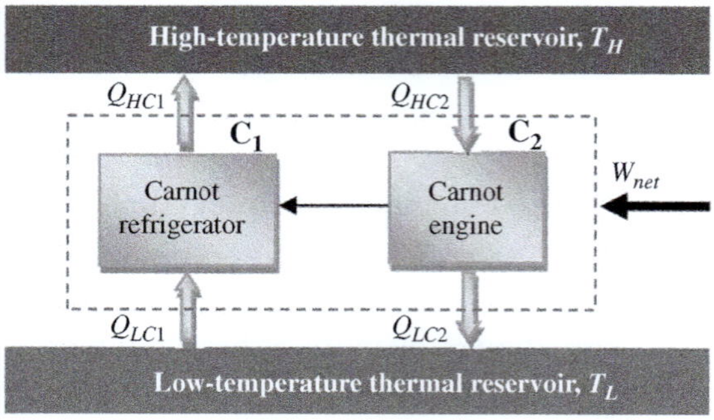

Fig. 3.2 Illustration of two Carnot engines operating between the same thermal reservoirs where $Q_{HC1} = Q_{HC2}$.

reverse direction as a refrigerator but C_2 remains to operate as a heat engine as depicted in Fig. 3.2. Because it is assumed that engine C_1 produces more work than engine C_2, it yields $Q_{LC1} < Q_{LC2}$. A combined system of refrigerator C_1 and engine C_2 (dashed rectangle area in Fig. 3.2) is a heat pump, which converts the entire work input $W_{net} = W_{C1} - W_{C2}$ to heat that is transferred to the low-temperature thermal reservoir.

$$W_{net} = W_{C_1} - W_{C_2} \tag{3.1}$$

$$Q_L = Q_{LC_2} - Q_{LC_1} \tag{3.2}$$

The combined system of Fig. 3.2 does not violate the second law. It is not in conflict with either statement of Kelvin-Planck or Clausius. The first law applied to the combined system shown in Fig. 3.2 gives

$$W_{net} = Q_L < 0 \tag{3.3}$$

Eq. (3.3) is applicable to heat pumps that interact with a single thermal reservoir only. Some authors interpret the second law by Eq. (3.3) for systems that communicate thermally with a single reservoir [2,3]. Because the combined system of Fig. 3.2 is not in violation of the second law, we conclude that between two Carnot engines operating between the same thermal reservoirs, it is possible that one may produce a higher work than the other, whereby invalidating Carnot' second corollary.

A further weakness of the proof of Carnot corollaries is that it only employs the possibility of the reverse operation of the Carnot engine—which is the first condition of the "perfect engine" as outlined by Carnot. The specifications of the perfect heat engine described by Carnot for production of a maximum amount of work from a given quantity of heat are [1]:

(1) if the same amount of work produced by the engine from a given amount of heat is spent to operate the engine backwards, an equal thermal effect will be produced,

(2) there should not be any temperature change, which may not be due to a volume change.

In Carnot's view, not necessarily any engine that could be operated in a reverse direction would produce a maximum work from a given amount of heat. Rather, both conditions mentioned above need to be fulfilled. The proof of Carnot corollaries given in Section 3.2.1 does not use the second criterion for maximum work production. So, the same reasoning applied to the Carnot engine in Fig. 3.1 may be used for any heat engine that could be operated in reverse direction with identical mechanical and thermal effects in both directions. For instance, an ideal Brayton cycle may be operated in a reverse direction, as a refrigerator, whereby the amount of work, heat input, and the rejected heat would be the same in both directions.

In the first series of operation, an amount of heat Q_H is supplied to an ideal Brayton engine to produce a network of W_{net}. The amount of heat rejected by the cycle is denoted by Q_L. In the reverse operation, by providing the same quantity of rejected heat Q_L and spending the same amount of work produced in the first series of operation, the quantity of heat rejected from the Brayton refrigerator would be identical to the quantity of heat supplied to the cycle in the first series of operation. So, one may apply the same reasoning of the proof of Carnot's corollaries to show that an ideal Brayton cycle is more efficient than any engine such as engine A in Fig. 3.1. However, as we know, the efficiency of ideal Brayton cycle is less than that of a Carnot cycle operating between the same fixed-temperature thermal reservoirs.

3.2.3 Carnot efficiency

The method of derivation of the Carnot efficiency presented in thermodynamics textbooks relies on Carnot's second principle, which is used to reason that the efficiency of the Carnot cycle depends solely on the temperatures of the thermal reservoirs. Although the method of demonstration of the corollaries suffers from certain issues, we now show that even the way the Carnot efficiency is derived is not entirely satisfactory.

Based on Carnot's second corollary, the efficiency of a Carnot cycle is reasoned to be a function of the reservoirs' temperatures, thus

$$\frac{Q_L}{Q_H} = f(T_H, T_L) \tag{3.4}$$

In the next step, three Carnot cycles operating between (T_H, T_L), (T_H, T_i), and (T_i, T_L), where $T_L < T_i < T_H$, are used to show that the function f can be represented as

$$f(T_H, T_L) = \frac{g(T_L)}{g(T_H)} \tag{3.5}$$

It follows that a combination of Eqs. (3.4) and (3.5) yields

$$\frac{Q_L}{Q_H} = \frac{g(T_L)}{g(T_H)} \tag{3.6}$$

where $g(T)$ is an unknown function of the temperature. In the next step, Eq. (3.6) is used to introduce the temperature scale, which was first proposed by William Thomson (Lord Kelvin) [4]; see the discussion of Section 2.3.5. Finally, $g(T) = T$ is selected as the simplest functional form for g whereby concluding $Q_L/Q_H = T_L/T_H$ and deriving the efficiency of the Carnot cycle, i.e. $\eta_C = 1 - T_L/T_H$. Some texts even skip the intermediate steps relating to Eqs. (3.5) and (3.6) and introduce the absolute temperature scale right after Eq. (3.4), e.g., see Ref. [2] Chapter 5 and Ref. [5] Chapter 5.

This method of derivation of the Carnot efficiency suffers from two issues. First, Eq. (3.4) that plays a major role in the derivation of the Carnot efficiency is valid subject to the validity of Carnot's second corollary. However, as discussed in Section 3.2.2, the proof of the corollaries is not concrete because certain objections can be raised. Second, the adoption of $g(T) = T$ appears to be rather a convention. Most authors state that Eq. (3.6) is the basis for determination of the absolute temperature scale.

In his 1848 paper [4], Thomson urged the need for a principle that would serve as a foundation for an absolute temperature scale. He noted that the temperature −273°C on the air-thermometer would correspond to the volume of air being reduced to nothing, a point of the scale which could not be reached at any finite temperature. Notice that the number 273 is the inverse of the expansion coefficient of air, i.e., 0.00366.

The law that provided a foundation to relate the two temperature scales (Kelvin and Centigrade) is the combined laws of Boyle-Mariotte and Dalton-Gay Lussac, which may be expressed as

$$pV = p_o V_o (1 + Et) \tag{3.7}$$

where p_o denotes the atmospheric pressure, V_o the volume at p_o and freezing temperature, E the expansion coefficient, t the temperature measured in degree centigrade, p and V represent the pressure and the volume of the gas at temperature t.

Substituting $T_o = \frac{1}{E}$ into Eq. (3.7) gives

$$pV = \frac{p_o V_o}{T_o} T \tag{3.8}$$

where $T_o = 273$ corresponds to the coefficient of thermal expansion of air and $T = 273 + t$.

Eq. (3.8) is the simplest state equation that is referred to as the ideal gas equation, which is used in the analytical arguments of Carnot [1], Clapeyron [6], Thomson [7], Rankine [8], and Clausius [9]. The numerical constant T_o employed in the analyses of Carnot and Clapeyron was 267, whereas it was 274.6 (inverse of the expansion coefficient of gases at the temperature of melting ice) in Rankine's work. Rankine also noted the absolute temperature of melting ice being 494.3 degrees in Fahrenheit scale.

3.2.4 Clausius inequality

The last step in the traditional method of teaching entropy is the presentation of the Clausius inequality given in Eq. (3.9), which then leads to the introduction of entropy as defined in Eq. (1.9). However, the Clausius inequality is usually introduced without sufficient background and without a clear connection to the previous steps. One would then need to figure out how to connect the dots, i.e., Carnot cycle, absolute temperature, reversibility, Clausius inequality, and entropy.

$$\oint \frac{\delta Q}{T} \leq 0 \tag{3.9}$$

In summary, the traditional method of introducing entropy as a thermodynamic property requires one to go through a lengthy and twisted process as schematically depicted in Fig. 3.3. The three intermediate steps in Fig. 3.3 are not fully justifiable as discussed in the previous sections. Furthermore, an introduction of the Clausius inequality as an analytical expression of the second law without a background is inappropriate. Among recent authors, Bejan [3] provides an analysis to derive the Clausius inequality. In the next section, a simple and straightforward approach is presented for introducing entropy in thermodynamics classes.

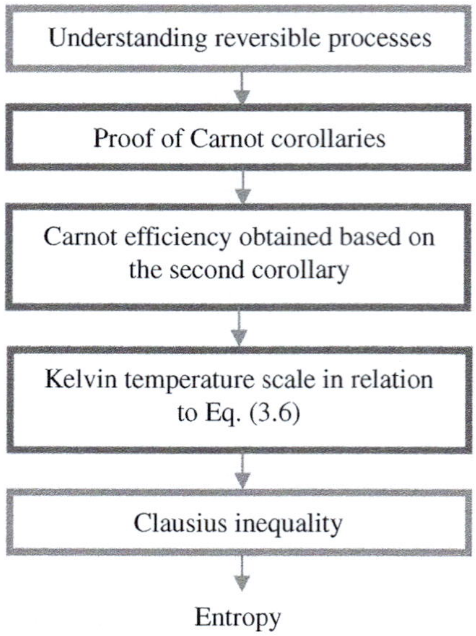

Fig. 3.3 The process of teaching entropy in thermodynamic classes.

3.3 A proposed method

The method to be proposed as an alternative to the present way of teaching entropy relies primarily on the original Clausius approach [9]. The first step after presenting the Clausius and Kelvin-Planck statements of the second law is to introduce the operation of the Carnot cycle on a p-V diagram and to derive its efficiency with an ideal gas as the working substance.

3.3.1 Derivation of the Carnot efficiency

As shown in Fig. 3.4, the Carnot cycle consists of adiabatic compression ($1 \rightarrow 2$), isothermal heating ($2 \rightarrow 3$), adiabatic expansion ($3 \rightarrow 4$), and isothermal cooling ($4 \rightarrow 1$). To derive an expression for the efficiency of the Carnot cycle, the amount of heat transferred to the cycle at temperature T_H and that rejected by the cycle to the low-temperature reservoir at temperature T_L are determined using the first law.

For the isothermal expansion of an ideal gas $dU=0$, the first law equation reduces to

$$\delta Q = \delta W \tag{3.10}$$

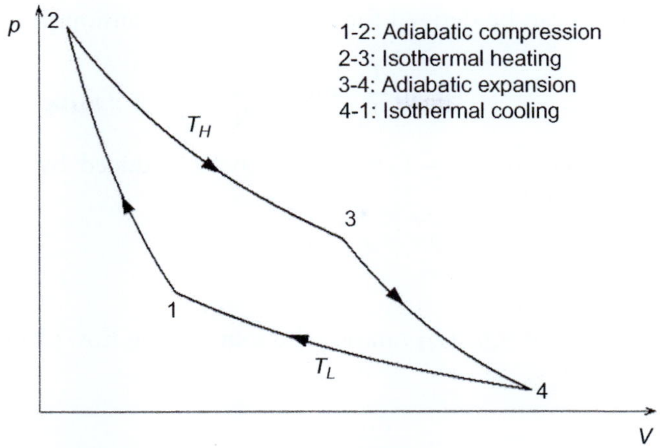

Fig. 3.4 The operation of the Carnot cycle on a *p-V* diagram.

The infinitesimal work done due to the expansion of a unit mass of the ideal gas is $\delta W = pdv$. Substituting $p = R_g T/v$, where v denotes the specific volume, and integrating Eq. (3.10) yields

$$Q_H = \int_2^3 R_g T \frac{dv}{v} = R_g T_H \ln \frac{v_3}{v_2} \tag{3.11}$$

where R_g is the gas constant. In a similar manner, the amount of heat rejected by the cycle per unit mole of the gas is found as follows.

$$Q_L = R_g T_L \ln \frac{v_4}{v_1} \tag{3.12}$$

For the adiabatic compression and expansion processes, the first law is

$$dU = -\delta W \tag{3.13}$$

Substituting $dU = c_v dT$ and $\delta W = pdv$ into Eq. (3.13) leads to

$$c_v dT = -pdv = -R_g T \frac{dv}{v} \tag{3.14}$$

Dividing both sides of Eq. (3.14) by T and integrating over the adiabatic compression process gives

$$c_v \ln \frac{T_2}{T_1} = R_g \ln \frac{v_1}{v_2} \tag{3.15}$$

A similar relation can be derived for the adiabatic expansion process.

$$c_v \ln \frac{T_4}{T_3} = R_g \ln \frac{v_3}{v_4} \tag{3.16}$$

Since $T_1 = T_4 = T_L$ and $T_2 = T_3 = T_H$, it can be deduced by comparing Eqs. (3.15) and (3.16) that

$$\frac{v_2}{v_1} = \frac{v_3}{v_4} \tag{3.17}$$

The heat transfer ratio Q_L/Q_H can now be found using Eqs. (3.11), (3.12), and (3.17). Hence,

$$\frac{Q_L}{Q_H} = \frac{T_L}{T_H} \tag{3.18}$$

Eq. (3.18) is valid for a Carnot cycle as well as the class of heat engines undergoing two isothermal processes, i.e., ideal Sterling and Ericson cycles.

Given the definition of the thermal efficiency of a heat engine as

$$\eta_{th} = \frac{\text{net work}}{\text{heat input}} = \frac{W_{net}}{Q_{in}} \tag{3.19}$$

where $W_{net} = Q_H - Q_L$, $Q_{in} = Q_H$, and using Eq. (3.18), the expression obtained for the Carnot efficiency is

$$\eta_C = 1 - \frac{T_L}{T_H} \tag{3.20}$$

3.3.2 Derivation of the Clausius integral

Eq. (3.18) is a remarkable result. For a Carnot cycle communicating with two thermal reservoirs, Eq. (3.18) can be rearranged to read

$$\frac{-Q_H}{T_H} + \frac{Q_L}{T_L} = 0 \tag{3.21}$$

Clausius called the second law the *theorem of the equivalence of transformations*. The first kind of transformation introduced by Clausius is the generation of heat at temperature T from work whose equivalence-value is Q/T. The second form of transformation is related to the transference of heat from a body at temperature T_1 to another at temperature T_2 [10]. In this case, the equivalence-value is

$$\frac{Q}{T_2} - \frac{Q}{T_1}$$

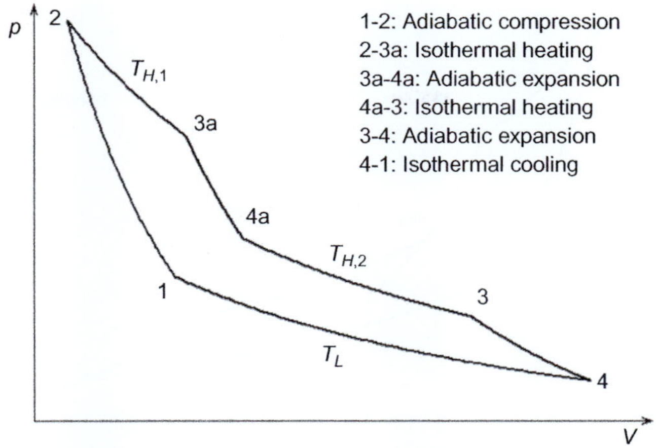

Fig. 3.5 The *p-V* diagram of a Carnot-like cycle with two isothermal heat addition and two adiabatic expansion processes.

The equivalence-value of the second type is indeed a combination of two equivalence-values of the first kind: transformation of heat Q at temperature T_1 into work and transformation of work into heat at temperature T_2.

Now, consider a Carnot-like cycle, which receives heat from two sources maintained at temperatures $T_{H,\ 1}$ and $T_{H,\ 2}$, but it rejects heat to a single low-temperature reservoir. Fig. 3.5 depicts the p-V diagram of the cycle. It consists of an adiabatic compression process 1(→2), two isothermal heating processes (2→3a and 4a→3), two adiabatic expansion processes (3a→4a and 3→4), and an isothermal cooling process (4→1).

If the heat received from the first and the second high-temperature reservoirs is denoted by $Q_{H,\ 1}$ and $Q_{H,\ 2}$, it can be shown that

$$\frac{-Q_{H,1}}{T_{H,1}}+\frac{-Q_{H,2}}{T_{H,2}}+\frac{Q_L}{T_L}=0 \tag{3.22}$$

To prove Eq. (3.22), the reversible cycle shown in Fig. 3.5 is divided into two Carnot cycles by extending the adiabatic expansion line 3a→4a whose intersection with the isothermal cooling line 4→1 is denoted by 4a′; see Fig. 3.6. The heat rejected by cycle I is $Q_{L,\ I}$ and that of cycle II is $Q_{L,\ II}$, where $Q_L = Q_{L,\ I} + Q_{L,\ II}$. Applying Eq. (3.21) to cycles I and II, one obtains

$$\frac{-Q_{H,1}}{T_{H,1}}+\frac{Q_{L,I}}{T_L}=0 \tag{3.23}$$

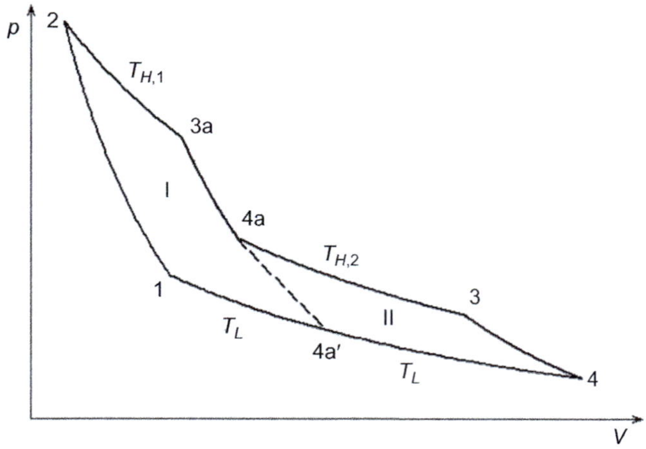

Fig. 3.6 Illustration of the cycle of Fig. 3.5 with two Carnot cycles designated as I and II.

$$\frac{-Q_{H,2}}{T_{H,2}} + \frac{Q_{L,II}}{T_L} = 0 \tag{3.24}$$

Adding Eqs. (3.23) and (3.24) leads to Eq. (3.22).

A similar analysis can be made for a Carnot-like cycle, which receives heat from a single heat source but rejects heat to two low-temperature reservoirs. As depicted in Fig. 3.7, the cycle consists of two adiabatic compression, two isothermal cooling, one isothermal heating, and one adiabatic

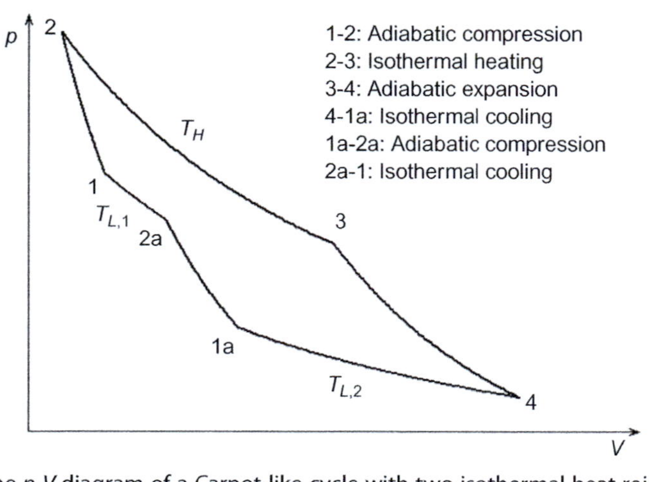

Fig. 3.7 The *p-V* diagram of a Carnot-like cycle with two isothermal heat rejection and two adiabatic compression processes.

expansion processes. If the heat rejected to the first and the second low-temperature reservoirs maintained at temperatures $T_{L,1}$ and $T_{L,2}$ is denoted by $Q_{L,1}$ and $Q_{L,2}$, respectively, one may obtain

$$\frac{-Q_H}{T_H} + \frac{Q_{L,1}}{T_{L,1}} + \frac{Q_{L,2}}{T_{L,2}} = 0 \tag{3.25}$$

Consider now a general case in that a reversible Carnot-like cycle communicates with several thermal reservoirs designated with temperatures T_1, T_2, T_3, …, T_{n-1}, T_n. The heat exchange between the cycle and the reservoirs is denoted by Q_1, Q_2, Q_3, …, Q_{n-1}, Q_n. It can be readily shown that the *algebraic* summation of all transformations Q_i/T_i is equal to zero.

$$\frac{Q_1}{T_1} + \frac{Q_2}{T_2} + \frac{Q_3}{T_3} + \ldots + \frac{Q_{n-1}}{T_{n-1}} + \frac{Q_n}{T_n} = 0 \tag{3.26}$$

Thus,

$$\sum_{i=1}^{n} \frac{Q_i}{T_i} = 0 \tag{3.27}$$

Eq. (3.27) is obtained assuming that the reservoirs are at constant temperatures. However, if the temperature changes during the heat exchange process, it is necessary to account for the variation of the temperature during the heat transfer. Thus, in general, it is more appropriate to show Eq. (3.27) in integral form as follows.

$$\oint \frac{\delta Q}{T} = 0 \tag{3.28}$$

where $\oint$ refers to cyclical integration.

3.3.3 Definition of entropy

The term under integral in Eq. (3.28) is the differential of a thermodynamic property to be denoted by S. It is called *entropy*, which is equivalent to *transformation* in Greek.

$$dS = \frac{\delta Q}{T} \tag{3.29}$$

As Clausius enunciated the second law as the theorem of the equivalence of transformations, he described entropy as the *transformational content* of a body.

It should be noted that in recent textbooks, Eqs. (3.28) and (3.29) often include the subscript "rev"; e.g., see Eq. (1.9), indicating that these relations are valid for reversible processes only. In general, when both reversible and irreversible processes are accounted for, Eq. (3.28) is expressed as

$$\Phi = \oint \frac{\delta Q}{T} \tag{3.30}$$

where Φ denotes *uncompensated transformation*, a terminology used by Clausius, which today is known as entropy generation. For irreversible processes $\Phi > 0$ and in the limit of reversible operation $\Phi = 0$.

3.3.4 Carnot cycle on *T-S* diagram

As shown in Fig. 3.4, the Carnot cycle consists of two reversible adiabatic and two reversible isothermal processes. Applying Eq. (3.29) to all four processes of the Carnot cycle and integrating over each process should allow one to establish relations between the entropies at the four corners of the cycle. For the reversible adiabatic compression process $1 \rightarrow 2$ where $\delta Q = 0$, one obtains $S_2 = S_1$. Likewise, for the reversible adiabatic expansion process, we have $S_3 = S_4$.

The heat exchange between the cycle and the reservoirs takes place during the isothermal processes only, which makes integration easier due to the temperature being constant. Integrating Eq. (3.29) over the isothermal heat addition process at temperature T_H yields

$$S_3 - S_2 = \frac{Q_H}{T_H} \tag{3.31}$$

Likewise, for the isothermal heat rejection process at temperature T_L, we have

$$S_4 - S_1 = \frac{Q_L}{T_L} \tag{3.32}$$

The operation of the Carnot cycle may now be shown on a new diagram with entropy and temperature as the abscissa and ordinate, respectively. From these results, it can be deduced that the *T-S* diagram of the Carnot cycle is a rectangle as depicted in Fig. 3.8.

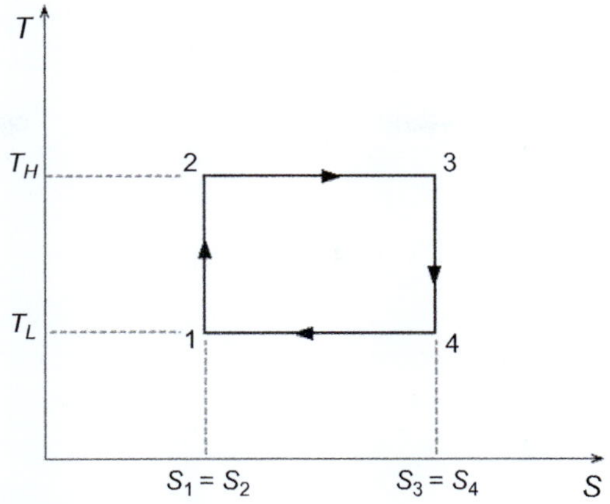

Fig. 3.8 Carnot cycle on a *T-S* diagram.

References

[1] S. Carnot, R.H. Thurston (Ed.), Reflections on the Motive Power of Heat, second ed., Wiley, New York, 1897.

[2] M.J. Moran, H.N. Shapiro, D.D. Boettner, M.B. Bailey, Fundamentals of Engineering Thermodynamics, ninth ed., Wiley, Hoboken, NJ, 2018.

[3] A. Bejan, Advanced Engineering Thermodynamics, fourth ed., Wiley, Hoboken, NJ, 2016.

[4] W. Thomson, On an absolute thermometric scale founded on Carnot's theory of the motive power of heat, and calculations from Regnault's observations, Philos. Mag. (1848) 100–106.

[5] C. Borgnakke, R.E. Sonntag, Fundamentals of Thermodynamics, ninth ed., Wiley, Hoboken, NJ, 2017.

[6] E. Clapeyron, Mémoire sur la puissance motrice de la chaleur, Journal de l'École Polytechnique 23 (1834) 153–190. https://gallica.bnf.fr/ark:/12148/bpt6k4336791.

[7] W. Thomson, An account of Carnot's theory of the motive power of heat; with numerical results deduced from Regnault's experiments on steam, Trans.Royal Soc. Edinb. XVI (1849) 113–155.

[8] W.J.M. Rankine, W.J. Millar (Eds.), Scientific Papers, Charles Griffin & Co., London, 1880.

[9] R. Clausius, The Mechanical Theory of Heat, Translated by W. R. Brown, MacMillan & Co., London, 1879.

[10] R. Clausius, in: T.A. Hirst (Ed.), The Mechanical Theory of Heat With Its Applications to the Steam-Engine and to the Physical Properties of Bodies, John van Voorst, London, 1867. 4th memoir.

CHAPTER FOUR

The common source of entropy increase

4.1 Introduction

Perhaps the entropy-heat relation was an obvious fact among the scientists of the 19th century. On page 162 of Maxwell's Theory of Heat [1], entropy is described as "when there is no communication of heat this quantity [entropy] remains constant, but when heat enters or leaves the body the quantity increases or diminishes." He further noted on page 194 that "entropy is a quantity such that without a change in its value no heat can enter or leave the body." In a 1907 thermodynamics textbook authored by Bryan [2], an English mathematician, a list of irreversible processes is given, which, as articulated by the author, can be categorized as the examples of irreversible conversion of work into heat. Friction between two rough surfaces, collision of two imperfectly elastic body, expansion of a gas into vacuum, and retardation of a fluid motion due to viscosity are some of the examples discussed by Bryan. The mechanical work that is lost in all these processes is converted into heat generated internally as explained by the author.

The proposition that we would like to put forward is that: Thermal effect is the only source of entropy generation. Alternatively, only irreversible processes that include a passage of heat may lead to a generation of entropy. It is important to realize that not necessarily every natural process may involve a flow of heat or thermal effect, e.g. aging, change in potential energy, moon orbiting around the earth, and earth around the sun. What we aim to accomplish in this chapter is to show that if processes like pressure drop, free expansion, and mixing of ideal gases lead to the production of entropy, it is due to the *thermal effect* involved in these processes.

Entropy Analysis in Thermal Engineering Systems
https://doi.org/10.1016/B978-0-12-819168-2.00004-0

4.2 Heat flow

The first and simplest example of irreversible process, which may lead to a production of entropy, is the flow of heat Q from a warmer body at temperature T_1 to a cooler body at temperature T_2. The two bodies may also represent two different locations in a single system. The change in the entropy of each body is obtained as follows.

$$\Delta S_1 = \int \frac{\delta Q}{T} = -\frac{Q}{T_1} \tag{4.1}$$

$$\Delta S_2 = \int \frac{\delta Q}{T} = -\frac{Q}{T_2} \tag{4.2}$$

The net change in the entropy is

$$\Phi = \frac{Q}{T_2} - \frac{Q}{T_1} \tag{4.3}$$

Because the warmer body loses heat, the change in its entropy is $-Q/T_1$, whereas the change in the entropy of the cooler body is Q/T_2 as it receives the same quantity of heat. Because $T_1 > T_2$, we conclude that $Q/T_2 > Q/T_1$. So, the net change in entropy (of both bodies) is positive. Thus, the heat flow is a source of entropy increase (generation).

In the example, each body is assumed to have a constant temperature during the heat transfer process. However, if the temperature varies, one must find a relation between the amount of heat and the temperature to perform integration in Eqs. (4.1) and (4.2). In this case, Eq. (4.3) is rewritten as follows.

$$\Phi = \int \left(\frac{1}{T_2} - \frac{1}{T_1} \right) \delta Q \tag{4.4}$$

4.3 Pressure drop

Entropy generation may take place due to pressure drop on the path of a fluid flow. The amount of work wasted due to the friction on the flow path is dissipated in the form of heat. A problem of this kind was analyzed by

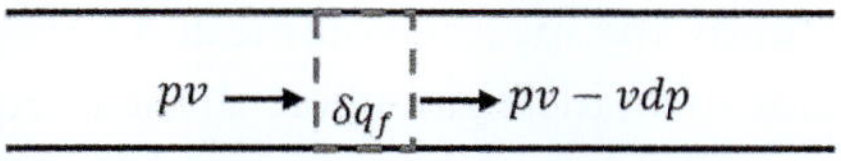

Fig. 4.1 Heat generation due to frictional work in a pipe flow.

Goodenough [3]. For an infinitesimal heat δq_f (per unit mass of the fluid) generated at temperature T, the increase of (specific) entropy is

$$ds = \frac{\delta q_f}{T} \tag{4.5}$$

Consider an incompressible fluid flowing through a horizontal pipe having a uniform cross-section as sketched in Fig. 4.1. The dashed rectangle shows a differential element where the temperature is T. The pipe is insulated so there is no heat exchange between the fluid and its surroundings. The work done on the left plane of the element is pv, whereas it is $pv - vdp$ on the right plane with p and v denoting the pressure and specific volume of the fluid, respectively. The amount of heat generated due to the friction equals the frictional work. Hence,

$$\delta q_f = (pv - v\delta p) - pv \tag{4.6}$$

Simplifying Eq. (4.6) yields

$$\delta q_f = -v\delta p \tag{4.7}$$

Eliminating dq_f between Eqs. (4.5) and (4.7), one obtains

$$ds = -\frac{v}{T}dp \tag{4.8}$$

The increase of entropy due to the pressure drop between two locations along the pipe is determined by integrating Eq. (4.8).

$$\Delta s = -v\int \frac{dp}{T} \tag{4.9}$$

Eq. (4.9) allows calculation of the entropy increase due to the *heat* generation associated with the pressure drop on the flow path of an incompressible fluid. Note that the analysis rests on the assumption of the entire heat generated being dissipated to the fluid.

For compressible fluids, the specific volume and pressure are interrelated. For example, in the case of an ideal gas where the state equation is $pv = R_g T$, Eq. (4.9) can be expressed as

$$\Delta s = -R_g \int \frac{dp}{p} \tag{4.10}$$

Note that $dp < 0$ in Eqs. (4.9) and (4.10); thus, $\Delta s > 0$. Because it was assumed that the frictional heat remains in the fluid, Δs denotes the increase in the entropy of the fluid.

4.4 Expansion

Let us now consider expansion of an ideal gas. In one arrangement, a rigid tank containing a pressurized gas is connected to another evacuated tank through a valve as shown in Fig. 4.2. If the valve is opened, the gas will pass from the tank to the one that is in vacuum. If we take the entire arrangement as a system, there is no work in this process, i.e., $W = 0$. The amount of heat exchange between the gas and its surroundings required to keep the gas temperature unaltered can be found by applying the first law. Hence,

$$Q = U_{final} - U_{initial} \tag{4.11}$$

Because the internal energy of an ideal gas depends on temperature only, it turns out that $U_{final} = U_{initial}$, so the net amount of heat exchange in the process is $Q = 0$.

Joule experimented air expansion using an apparatus like that shown in Fig. 4.2. One tank was filled with air and it was compressed to 22 atm, which was then connected to another evacuated tank. The two tanks were then placed in a water calorimeter. After expansion of the air to about twice its initial volume, no loss of heat was observed from the calorimeter [4].

In another experiment, Joule placed the two tanks and the connecting pipe in three different calorimeters. It was observed that the tank containing

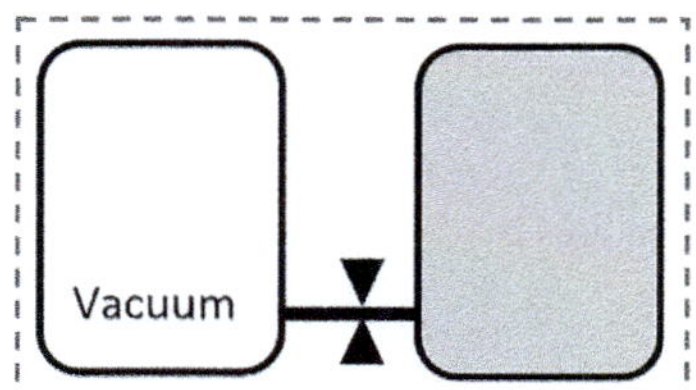

Fig. 4.2 Expansion of an ideal gas into vacuum.

the air gained heat, whereas the pipe and the second tank lost heat. Interesting to note is that the heat gained was almost equal to the total heat lost. To correctly determine entropy generation, one would need to account for the heat transfer processes taking place between the different parts of the apparatus and the calorimeters.

Consider now a rigid tank, which, as before, contains a compressed ideal gas. It is then allowed to escape to the atmosphere. An expansion like this with air as the gaseous medium was also experimented by Joule [4]. The escaping air passed through a long coil of pipe in order to reduce the temperature of the air to its initial temperature. The entire assembly was placed in a water calorimeter. Joule's observation was that heat was given off to the expanding air yielding a cooling effect in the calorimeter. These experiments led Joule to determine the mechanical equivalent of heat. The external work performed by the escaping air to overcome the atmospheric pressure was found to be the same as the amount of heat absorbed by the air. The analytical formulation of Joule's experiment was presented by Thomson [5] and later Clausius [6]. They both concluded that in the expansion process of air (treated as an ideal gas), it is necessary to supply heat to maintain a constant temperature. From these arguments, it should be obvious now that the entropy increase associated with the isothermal expansion of an ideal gas is indeed due to the heat transferred to the gas from an external source.

For an infinitesimal amount of work done in the expansion of an ideal gas, we have $\delta Q = \delta W$, where δQ denotes the infinitesimal amount of heat supplied to maintain an isothermal expansion of the gas. Substituting $\delta Q = TdS$ and $\delta W = pdV$ gives

$$TdS = pdV \tag{4.12}$$

Integrating Eq. (4.12) from an initial volume to a final volume with the use of the ideal gas equation, one obtains

$$\Delta S = nR \ln\left(\frac{V_f}{V_i}\right) \tag{4.13}$$

where ΔS is the increase in the entropy of the expanding gas.

Assume that the total heat Q required in the expansion process is supplied from a source whose temperature during the infinitesimal heat transfer δQ is T_s. The net decrease in the entropy of the source is

$$\Delta S_s = -\int \frac{\delta Q}{T_s} \tag{4.14}$$

The negative sign indicates that the heat is lost from the source. The total entropy generation associated with isothermal expansion of an ideal gas is obtained as follows.

$$\Phi = nR \ln\left(\frac{V_f}{V_i}\right) - \int \frac{\delta Q}{T_s} \tag{4.15}$$

4.5 Mixing

A further widely known process that leads to a production of entropy is mixing two or more fluids. We will first consider mixing of liquids and then gases. Fig. 4.3 shows two fluids entering an insulated chamber where they mix and then leave at the exit port 3. Assume that the two fluids entering the chamber through ports 1 and 2 are identical liquids having the same temperature and pressure. As the process is adiabatic and $T_1 = T_2$, it can be deduced from the first law that the temperature at the exit port T_3 is the same as the inlet temperature, i.e., $T_3 = T_2 = T_1$. Because state 3 is identical to states 1 and 2, the entropy of the mixed stream leaving the chamber is also the same as that at the port of entries. The process does not involve any form of heat, nor entropy is generated.

Next, consider two mixing liquids, not necessarily identical, entering the chamber with different temperatures such that $T_2 > T_1$. Within the chamber, heat is transferred from the warmer fluid to the cooler so the mixed stream leaves at T_3, where $T_1 < T_3 < T_2$. For an infinitesimal heat exchange δQ between the two fluids, the entropy production is determined as

$$d\Phi = \frac{\delta Q}{T_1'} - \frac{\delta Q}{T_2'} \tag{4.16}$$

where T_1' and T_2' denote the temperatures of the two fluids during the infinitesimal heat transfer δQ.

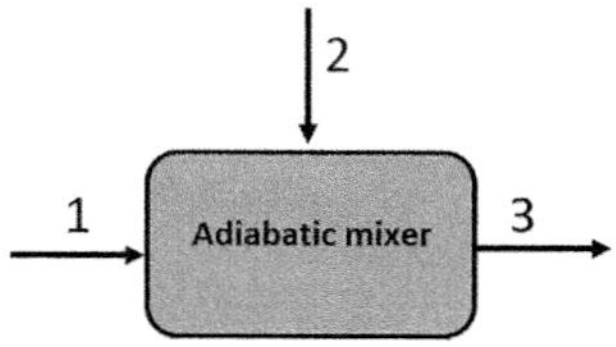

Fig. 4.3 Mixing of two liquids in a perfectly insulated chamber.

Noting that $\delta Q = m_1 c_1 dT_1' = m_2 c_2 dT_2'$ where c denotes the specific heat, the total entropy generation is obtained by integrating Eq. (4.16). Hence,

$$\begin{aligned}\Phi &= m_1 c_1 \int_{T_1}^{T_3} \frac{dT_1'}{T_1'} - m_2 c_2 \int_{T_2}^{T_3} \frac{dT_2'}{T_2'} \\ &= m_1 c_1 \ln \frac{T_3}{T_1} + m_2 c_2 \ln \frac{T_2}{T_3}\end{aligned} \tag{4.17}$$

Consider now mixing process of ideal gases. For simplicity of the analysis, assume the gases have identical temperature and pressure and that mixing takes place at a uniform temperature and pressure. A problem of this kind was presented by Dodge [7] in 1944. The pressure of each gaseous species then decreases from an initial pressure p_i to its partial pressure p_m in the mixture. It can therefore be stated that in the process of mixing of ideal gases, each expands from the initial pressure down to its partial pressure in the mixture. Succinctly, the isothermal mixing of ideal gases is equivalent to the isothermal expansion of its individual species.

The expansion work of each gas is obtained as follows.

$$W_g = \int p dV = nRT \ln \frac{p_i}{p_m} \tag{4.18}$$

We discussed in Section 4.4 that the isothermal expansion of an ideal gas requires heat. Applying the first law to an ideal gas undergoing an isothermal expansion ($\Delta U = 0$) yields $W_g = Q$. The expansion of each ideal gas during the isothermal mixing process would leave a cooling effect on its immediate surroundings, which is the reason for the entropy increase of the individual gases, i.e., $\Delta S_g = Q/T = W_g/T$. For the ideal gases being mixed in an insulated system, the heat exchange is expected to take place between the mixing gases. Dividing Eq. (4.18) by the temperature gives the entropy increase associated with the isothermal expansion of each gas.

$$\Delta S_g = nR \ln \frac{p_i}{p_m} \tag{4.19}$$

Eq. (4.19) is in principle equivalent to Eq. (4.13). The total entropy increase associated with the mixing of k ideal gases is obtained as

$$\Delta S_{mix} = \sum^{k} \Delta S_g \tag{4.20}$$

Fig. 4.4 Mixing equimolar ideal gases in a perfectly insulated chamber.

Consider now isothermal mixing of equimolar ideal gases that initially occupy an identical volume V_i in a perfectly insulated chamber as shown in Fig. 4.4. From the state equation $pV=nRT$, it can be deduced that all gases have also an identical pressure p_i before mixing. The state equation for the mixture can be expressed as $p_{mix}V_{mix}=n_{mix}RT$, where

$$V_{mix}=\sum_{j=1}^{k}V_j=kV_i \tag{4.21}$$

$$n_{mix}=\sum_{j=1}^{k}n_j=kn_i \tag{4.22}$$

Substituting Eqs. (4.21) and (4.22) into the state equation we find $p_{mix}=p_i$; that is, the pressure of the mixture is the same as the pressure of the gases before mixing. Because the volume of each gas increases from V_i to V_{mix}, the pressure of the gas decreases from p_i to p_i/k. The increase in the entropy is obtained using Eqs. (4.19) and (4.20).

$$\Delta S_{mix}=\sum_{j=1}^{k}\left(nR\ln\frac{p_i}{\frac{1}{k}p_i}\right)_j=\sum_{j=1}^{k}(nR\ln k)_j=n_iRk\ln k \tag{4.23}$$

4.6 Interpretation of entropy

Our investigation of the irreversible phenomena discussed in the preceding sections reveals that the presence of heat is the sole reason for the entropy increase in these processes confirming the preposition put forward in Section 4.1:

Entropy generation may take place only in irreversible processes that include a passage of heat.

Pressure drop and expansion of ideal gases are common examples of irreversible conversion of work into heat. Recall the general entropy balance equation, Eq. (1.16), which includes heat-to-temperature ratio and entropy (specific entropy multiplied by mass) terms where the entropy of a substance is determined using Eq. (1.11). Thus, there should not be a surprise that the only phenomenon responsible for entropy generation is heat transfer. A claim that entropy generation can take place in the absence of heat would be in contradiction of the original definition of entropy, Eq. (1.9). A change in the entropy of a body strictly depends on the amount of heat and its temperature at which the heat transfer takes place. Entropy production is an indication of the extent of heat transfer.

Several interpretations have been presented for entropy, which have no or very limited connection to its original definition as introduced by Clausius. For instance, entropy has been viewed as a measure of disorder. This notion is originated from Boltzmannian definition of entropy that relies on the molecular motion or the theory of probabilities where entropy is described as a product of k_B, Boltzmann constant, and ln W where W denotes the number of microstates. The relation for the Boltzmannian entropy $S = k_B \ln W$ rests on a critical assumption that the entropy of a state depends on the probability of that state [8]. The notion of entropy as a measure of disorder has been discredited by recent scholars [9,10].

Entropy has also been interpreted as a measure of energy dispersal. In 1910, Kline wrote: "Growth of entropy is a passage from a concentrated to a distributed condition of energy. Energy originally concentrated variously in the system is finally scattered uniformly in said system" [8]. This view of entropy was further elucidated in 1996 by Leff [11] and a few years later by Lambert [10,12]. The energy dispersal view of entropy, undoubtfully better than disorder, is also based upon the statistical definition of entropy, i.e., $k_B \ln W$, without regard to its original definition. For instance, the entropy increase associated with the expansion of a gas is explained to be due to microstates being more accessible in the final state than the initial because the volume increases. The reasoning given for the case of mixing of two ideal gases in a box, which are initially separated by a partition, is similar, but the thermal effect is overlooked.

It must be noted that the reason why entropy generation is usually related to mere irreversibility of expansion and mixing processes is the confusion that appears to exist with respect to the combined first and second laws equation, Eq. (1.19), and that the Clausius definition of entropy is replaced with statistical arguments, which rely on $S = k_B \ln W$. The common misperception with respect to Eq. (1.19) is that it is usually treated as a self-standing thermodynamic equation without recognizing where it is originated from. Eq. (1.19) is an alternative expression of the first law equation

$\delta Q - \delta W = dU$ with $\delta Q = TdS$ and $\delta W = pdV$. For an ideal gas undergoing an isothermal process, the first law is expressed as $\delta Q = \delta W$ so Eq. (1.19) reduces to $TdS = pdV$. In the absence of any work interaction (both external and internal), Eq. (1.19) becomes $dU = TdS$. If the process does not include any heat (supplied externally or generated internally), the process is therefore isentropic and Eq. (1.19) reduces to $dU = pdV$. Finally, for a system that experiences no form of work and heat, the change in the internal energy is zero $dU = 0$.

The phenomenon of entropy increase (generation) is not merely correlated with the irreversibility of the natural processes. It is rather an indicative measure of the transference of energy in the form of heat, which may be supplied from an external source, be generated internally by conversion of work to heat, or be exchanged internally within a system. Interpreting entropy generation as a measure of heat transfer may be considered as a more accurate form of the energy dispersal view. The former takes into account both the amount of heat and the temperature, whereas the latter accounts only for spreading of energy without specifying the form of energy.

References

[1] J.C. Maxwell, Theory of Heat, Longmans Green & Co., London, 1902 pp. 162, 194.
[2] G.H. Bryan, Thermodynamics: An Introductory Treatise Dealing Mainly With First Principles and Their Direct Applications, B. G. Teubner, Leipzig, 1907, pp. 71–74.
[3] G.A. Goodenough, Principles of Thermodynamics, second ed., Henry Hold & Co., New York, NY, 1914.
[4] J.P. Joule, On the changes of temperature produced by the rarefaction and condensation of air, Philos. Mag. 26 (174) (1845) Ser. 3.
[5] W. Thomson, An account of Carnot's theory of the motive power of heat; with numerical results deduced from Regnault's experiments on steam, Trans. Royal Soc. Edinb., Vo. XVI, Part I, 541–574.
[6] R. Clausius, The Mechanical Theory of Heat, Translated by W. R. Brown, MacMillan & Co., London, 1879.
[7] B.F. Dodge, Chemical Engineering Thermodynamics, McGraw-Hill, New York, 1944, pp. 114–116.
[8] J.F. Klein, Physical Significance of Entropy or of the Second Law, Van Nostrand Company, (1910).
[9] A. Bejan, Thermodynamics today, Energy 160 (2018) 1208–1219.
[10] F.L. Lambert, Shuffled cards, messy desks, and disorderly dorm rooms—examples of entropy increase? Nonsense, J. Chem. Educ. 76 (1999) 1385.
[11] H.S. Leff, Thermodynamic entropy: the spreading and sharing of energy, Am. J. Phys. 64 (1996) 1261–1271.
[12] F.L. Lambert, Disorder—a cracked crutch for supporting entropy discussions, J. Chem. Educ. 79 (2002) 187–192.

CHAPTER FIVE

Most efficient engine

5.1 Introduction

Heat engines have played a significant role in modernization of the mankind's life. Early engines had a very low efficiency, so it was one of the challenges of engineers to find new methods and designs to increase the power production efficiency. Sadi Carnot, a French military engineer, was determined to answer a central question of his ear: "whether the motive power of heat is unbounded," and "whether the possible improvements in steam engines have an assignable limit, a limit which the nature of the things will not allow to be passed by any means whatever" [1].

In his investigation to determine an upper limit for the efficiency of heat engines, Carnot had considered certain design constraints: (i) the quantity of heat is given, and (ii) the highest and lowest temperatures experienced by engine are fixed. In his era, neither the first law nor the second had fully been realized and formulated. His analysis was based on (i) the caloric theory where heat was sought to be an indestructible substance, which would transfer between two bodies with different temperatures, and (ii) the empirical laws of Boyle-Mariotte and Dalton-Gay Lussac.

Carnot's investigation led him to propose a design of ideal engine. Since then, the Carnot cycle has been used as a reference to measure the effectiveness of other engines. In the comparison of the performance of an engine with that of a Carnot cycle, it is traditionally assumed that both engines operate between the same high- and low-temperature thermal reservoirs—the second design constraint of Carnot. Under this condition, a class of heat engines with two isothermal processes (i.e., Stirling, Carnot, and Ericsson engines) are the most efficient engines. The reason for this is adequately given by Rankine [2]:

> *As the conversion of heat into expansive power arises from changes of volume only, and not from changes of temperature, it is obvious, that the proportion of the heat received which is converted into expansive power will lie the greatest possible, when the reception of heat, and its emission, each take place at a constant temperature.*

Entropy Analysis in Thermal Engineering Systems
https://doi.org/10.1016/B978-0-12-819168-2.00005-2

It is natural to be also curious about the design of an ideal engine when all engines are constrained to undergo an identical degree of volume change. In the following sections, we will derive and compare expressions for the efficiency of common thermodynamic power cycles. The efficiency comparison will be made under the assumption that the largest and smallest volumes experienced by the working medium are the same (i.e., design constraint) in all engines.

5.2 Thermodynamic power cycles

A list of common power cycles is provided in Table 5.1 along with the inventor, year, and place of invention for each design. Evident from Table 5.1 is that all these engines (except the Miller cycle) were invented throughout the 19th century. There are other engine designs such as Rankine cycle, Lenoir cycle, Dual cycle, and Stoddard engine as well as modern designs such as Allam and Kalina power cycles [3, 4], but they will not be discussed in this chapter.

In the following sections, an expression will be derived for the thermal efficiency of the cycles given in Table 5.1. It will be assumed that the working fluid is an ideal gas with a constant specific heat throughout the cycle. The idea is to describe the engine efficiency in terms of the compression ratio (CR) and pressure ratio (PR).

$$\eta = f(CR, PR) \tag{5.1}$$

where

$$CR = V_{max}/V_{min} \tag{5.2}$$

Table 5.1 A list of common gas power cycles.

Cycle	Inventor	Year of invention	Place of invention
Stirling	Robert Stirling	1816	United Kingdom
Carnot	Nicolas Léonard Sadi Carnot	1824	France
Ericsson	John Ericsson	1853	United States
Brayton	George Bailey Brayton	1872	United States
Otto	Nikolaus August Otto	1876	Germany
Atkinson	James Atkinson	1882	United Kingdom
Diesel	Rudolf Christian Karl Diesel	1893	Germany
Miller	Ralph Miller	1957	United States

$$PR = p_{max}/p_{min} \tag{5.3}$$

Indeed, CR denotes the maximum degree of volume change experienced by the working gas, whereas PR is the highest degree of the pressure change that the gaseous medium undergoes in each cyclical operation.

5.2.1 Stirling cycle

A p-V diagram of the Stirling cycle is depicted in Fig. 5.1. It consists of four processes: isothermal compression with heat removal $1 \rightarrow 2$, isochoric heat addition $2 \rightarrow 3$, isothermal expansion with heat addition $3 \rightarrow 4$, and isochoric heat removal $4 \rightarrow 1$. Thus, we have $T_1 = T_2$, $V_2 = V_3$, $T_3 = T_4$, and $V_4 = V_1$. The thermal efficiency of the Stirling cycle is the same as the Carnot efficiency. Hence,

$$\eta = 1 - \frac{T_1}{T_3} \tag{5.4}$$

Using the relation $p_1V_1/T_1 = p_3V_3/T_3$, Eq. (5.4) can be expressed as

$$\eta = 1 - \frac{p_1V_1}{p_3V_3} \tag{5.5}$$

The compression and pressure ratios for the Stirling cycle are $CR = V_1/V_3$ and $PR = p_3/p_1$. Eq. (5.5) may now be represented in terms of CR and PR.

$$\eta = 1 - \frac{CR}{PR} \tag{5.6}$$

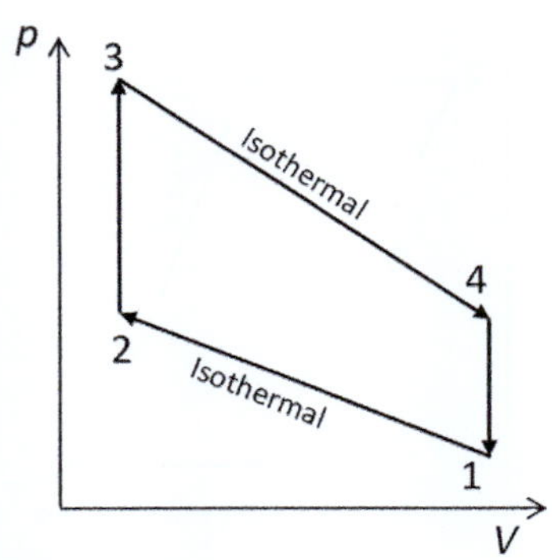

Fig. 5.1 A p-V diagram of the Stirling cycle.

Denote $T_R = T_{max}/T_{min}$ as the ratio of the maximum-to-minimum temperature of the cycle, where $T_{max} = T_3$ and $T_{min} = T_1$, it can be inferred from Eqs. (5.4) and (5.6) that $PR = CR \cdot T_R$.

5.2.2 Brayton cycle

The Brayton cycle consists of the following processes: adiabatic compression $1 \to 2$, isobaric heat addition $2 \to 3$, adiabatic expansion $3 \to 4$, and isobaric heat removal. A p-V diagram of the cycle is depicted in Fig. 5.2. For the Brayton cycle, we have $p_2 = p_3$, $p_4 = p_1$, $CR = V_4/V_2$, and $PR = p_2/p_4$. The thermal efficiency of the cycle obeys [5]

$$\eta = 1 - \frac{T_1}{T_2} \tag{5.7}$$

For the adiabatic compression process $1 \to 2$, we have

$$\frac{T_1}{T_2} = \left(\frac{p_1}{p_2}\right)^{1-\frac{1}{\gamma}} = \left(\frac{p_4}{p_2}\right)^{1-\frac{1}{\gamma}} = PR^{\frac{1}{\gamma}-1} \tag{5.8}$$

Substituting Eq. (5.8) into Eq. (5.7) yields

$$\eta = 1 - PR^{\frac{1}{\gamma}-1} \tag{5.9}$$

A relationship can be established between CR and PR in the Brayton cycle. Using the relation $\frac{p_2 V_2}{T_2} = \frac{p_4 V_4}{T_4}$, we write

$$CR = PR\frac{T_4}{T_2} \tag{5.10}$$

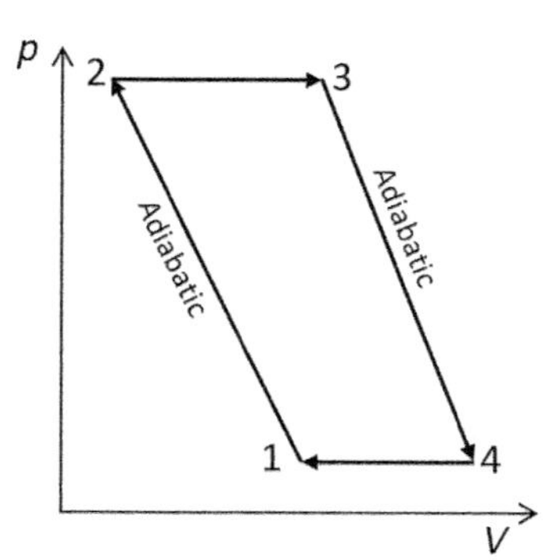

Fig. 5.2 A p-V diagram of the Brayton cycle.

From Eq. (5.8), we have $T_2 = T_1 PR^{1-\frac{1}{\gamma}}$. Likewise, for the adiabatic expansion process 3→4, we get $T_4 = T_3 PR^{\frac{1}{\gamma}-1}$. Substituting these two relations into Eq. (5.10) gives

$$CR = T_R PR^{\frac{2}{\gamma}-1} \tag{5.11}$$

where $T_R = T_3/T_1$.

Eliminating PR between Eqs. (5.9) and (5.11) leads to an alternative expression for the efficiency of the Brayton cycle.

$$\eta = 1 - \left(\frac{CR}{T_R}\right)^{\frac{1-\gamma}{2-\gamma}} \tag{5.12}$$

5.2.3 Otto cycle

The operation of the Otto cycle on a p-V diagram is shown in Fig. 5.3. The cycle comprises four processes: adiabatic compression 1→2, isochoric heat addition 2→3, adiabatic expansion 3→4, and isochoric heat removal 4→1. Thus, for the Otto cycle, we have $V_2 = V_3$, $V_4 = V_1$, $CR = V_1/V_3$, and $PR = p_3/p_1$. The thermal efficiency of the Otto cycle is [5]

$$\eta = 1 - \frac{T_1}{T_2} \tag{5.13}$$

Applying the first law to the adiabatic compression process 1→2 gives $\frac{T_1}{T_2} = \left(\frac{V_1}{V_2}\right)^{1-\gamma}$. Thus, Eq. (5.13) may be rewritten as

$$\eta = 1 - \left(\frac{V_1}{V_2}\right)^{1-\gamma} = 1 - CR^{1-\gamma} \tag{5.14}$$

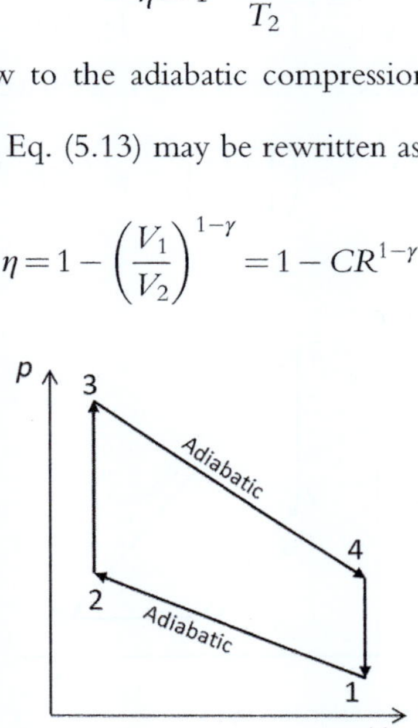

Fig. 5.3 A p-V diagram of the Otto cycle.

For the Otto cycle, $T_{max}=T_3$ and $T_{min}=T_1$. So, from the relation $\frac{p_1 V_1}{T_1}=\frac{p_3 V_3}{T_3}$, one finds $PR=CR \cdot T_R$.

5.2.4 Atkinson cycle

The operation of the Atkinson cycle is depicted on a p-V diagram in Fig. 5.4. The cycle comprises the following processes: adiabatic compression $1 \rightarrow 2$, isochoric heat addition $2 \rightarrow 3$, adiabatic expansion $3 \rightarrow 4$, and isobaric heat removal. For this cycle, we have $V_2=V_3$, $p_4=p_1$, $CR=V_4/V_3$, $PR=p_3/p_4$, and $T_R=T_3/T_1$.

The amount of heat supplied during the process $2 \rightarrow 3$ to a unit mass of the working gas is

$$q_{23}=c_v(T_3-T_2) \tag{5.15}$$

The amount of heat removed during the process $4 \rightarrow 1$ from a unit mass of the gas is

$$q_{41}=c_p(T_4-T_1) \tag{5.16}$$

Now, we take the ratio of the two quantities of heat in Eqs. (5.15) and (5.16) as follows

$$\frac{q_{41}}{q_{23}}=\gamma \frac{(T_4/T_1)-1}{(T_3/T_1)-(T_2/T_1)} \tag{5.17}$$

For the adiabatic processes $1 \rightarrow 2$ and $3 \rightarrow 4$, we have

$$\frac{T_2}{T_1}=\left(\frac{V_2}{V_1}\right)^{1-\gamma}=\left(\frac{V_3}{V_4}\frac{T_4}{T_1}\right)^{1-\gamma}=\left(CR^{-1}\frac{T_4}{T_1}\right)^{1-\gamma} \tag{5.18}$$

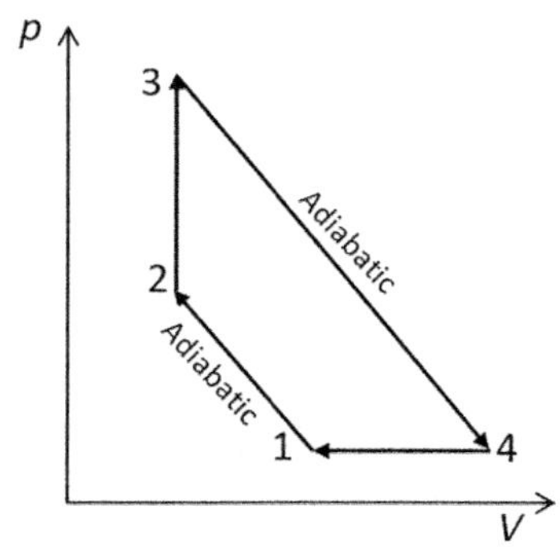

Fig. 5.4 A p-V diagram of the Atkinson cycle.

$$\frac{T_4}{T_3}=\left(\frac{V_4}{V_3}\right)^{1-\gamma}=CR^{1-\gamma} \tag{5.19}$$

Eq. (5.19) may be employed to determine an expression for (T_4/T_1). Hence,

$$\frac{T_4}{T_1}=T_R CR^{1-\gamma} \tag{5.20}$$

A substitution of Eq. (5.20) into Eq. (5.18) yields

$$\frac{T_2}{T_1}=\left(T_R CR^{-\gamma}\right)^{1-\gamma} \tag{5.21}$$

The thermal efficiency of the cycle may now be obtained upon substituting Eqs. (5.20) and (5.21) into Eq. (5.17). Hence,

$$\eta=1-\gamma\frac{T_R CR^{1-\gamma}-1}{T_R-\left(T_R CR^{-\gamma}\right)^{1-\gamma}} \tag{5.22}$$

Note that the relation between CR and PR in the Atkinson cycle obeys $CR=PR^{\frac{1}{\gamma}}$. Thus, an alternative expression for the thermal efficiency is

$$\eta=1-\gamma\frac{T_R PR^{\frac{1}{\gamma}-1}-1}{T_R-\left(\frac{T_R}{PR}\right)^{1-\gamma}} \tag{5.23}$$

5.2.5 Diesel cycle

Fig. 5.5 shows a p-V diagram of the Diesel cycle, which consists of the following four processes: adiabatic compression $1\rightarrow2$, isobaric heat addition

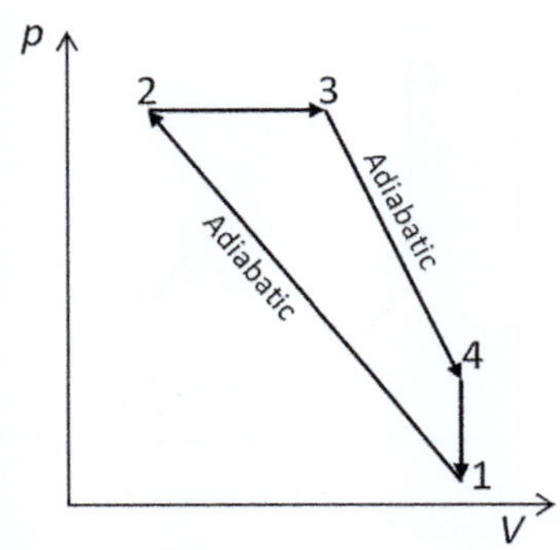

Fig. 5.5 A *p-V* diagram of the Diesel cycle.

$2 \rightarrow 3$, adiabatic expansion $3 \rightarrow 4$, and isochoric heat removal. For this cycle, we have $p_2 = p_3$, $V_4 = V_1$, $CR = V_1/V_2$, $PR = p_2/p_1$, and $T_R = T_3/T_1$. The thermal efficiency of the cycle can be described as

$$\eta = 1 - \frac{q_{41}}{q_{23}} = 1 - \frac{1}{\gamma}\left(\frac{T_4/T_1 - 1}{T_3/T_1 - T_2/T_1}\right) \tag{5.24}$$

For the adiabatic processes $1 \rightarrow 2$ and $3 \rightarrow 4$, we have

$$\frac{T_2}{T_1} = \left(\frac{V_2}{V_1}\right)^{1-\gamma} = CR^{\gamma-1} \tag{5.25}$$

$$\frac{T_4}{T_3} = \left(\frac{V_4}{V_3}\right)^{1-\gamma} = \left(\frac{V_1}{V_2}\frac{T_2}{T_3}\right)^{1-\gamma} = \left(\frac{CR^{\gamma}}{T_R}\right)^{1-\gamma} \tag{5.26}$$

Note that Eq. (5.25) and the relation $V_3/V_2 = T_3/T_2$ are used in Eq. (5.26). The thermal efficiency of the Diesel cycle can now be obtained in terms of T_R and CR using Eqs. (5.25) and (5.26) in Eq. (5.24). Hence,

$$\eta = 1 - \frac{1}{\gamma}\left[\frac{T_R^{\gamma}(CR^{\gamma})^{1-\gamma} - 1}{T_R - CR^{\gamma-1}}\right] \tag{5.27}$$

Like the Atkinson cycle, the relation between CR and PR obeys $CR = PR^{\frac{1}{\gamma}}$.

5.2.6 Miller cycle

The operation of the Miller cycle is shown on a p-V diagram in Fig. 5.6. It includes five processes: adiabatic compression $1 \rightarrow 2$, isochoric heat addition $2 \rightarrow 3$, adiabatic expansion $3 \rightarrow 4$, isochoric heat removal $4 \rightarrow 5$, and isobaric

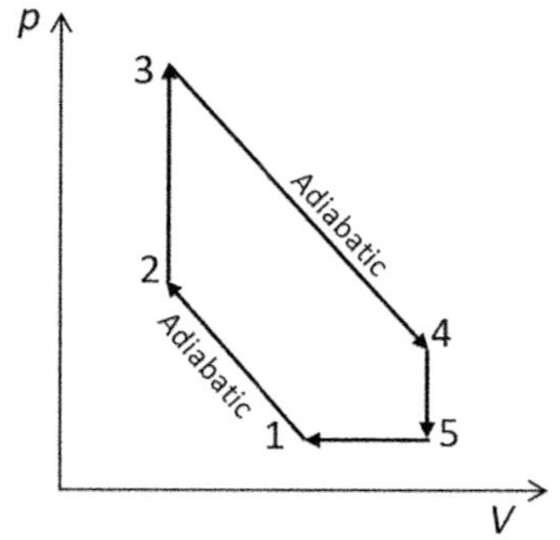

Fig. 5.6 A *p*-*V* diagram of the Miller cycle.

heat removal $5 \rightarrow 1$. For the Miller cycle, we have $V_2 = V_3$, $V_4 = V_5$, $p_5 = p_1$, $CR = V_5/V_3$, $PR = p_3/p_5$, and $T_R = T_3/T_1$. The thermal efficiency of the cycle is determined as follows.

$$\eta = 1 - \frac{q_{45} + q_{51}}{q_{23}} = 1 - \frac{c_v(T_4 - T_5) + c_p(T_5 - T_1)}{c_v(T_3 - T_2)} \tag{5.28}$$

Upon introducing the specific heat ratio, Eq. (5.28) may be rearranged to read

$$\eta = 1 - \frac{(T_4/T_1) + (\gamma - 1)(T_5/T_1) - \gamma}{T_R - (T_2/T_1)} \tag{5.29}$$

The temperature ratios (T_4/T_1), (T_5/T_1), and (T_2/T_1) in Eq. (5.29) can be determined using the relations of the adiabatic processes. For the adiabatic expansion process $3 \rightarrow 4$, we have

$$\frac{T_4}{T_3} = \left(\frac{V_4}{V_3}\right)^{1-\gamma} = \left(\frac{V_5}{V_3}\right)^{1-\gamma} = CR^{1-\gamma} \tag{5.30}$$

and

$$p_4 = p_3\left(\frac{V_4}{V_3}\right)^{-\gamma} = p_3 CR^{-\gamma} \tag{5.31}$$

From Eq. (5.30), we find the following expression for T_4/T_1.

$$\frac{T_4}{T_1} = T_R CR^{1-\gamma} \tag{5.32}$$

To obtain a relation for T_5/T_1, we use Eq. (5.32) and the relation $T_5 = T_4(p_5/p_4)$ that is applicable to the isochoric process $4 \rightarrow 5$.

$$\frac{T_5}{T_1} = \frac{T_4 p_5}{T_1 p_4} = \left(T_R CR^{1-\gamma}\right)\frac{p_5}{p_3 CR^{-\gamma}} = T_R\frac{CR}{PR} \tag{5.33}$$

The last temperature ratio in Eq. (5.29) is determined using the relationship that is valid for the adiabatic compression process $1 \rightarrow 2$, and $V_1 = V_5(T_1/T_5)$ applicable to the isobaric process $5 \rightarrow 1$. Hence,

$$\frac{T_2}{T_1} = \left(\frac{V_2}{V_1}\right)^{1-\gamma} = \left(\frac{V_3}{V_5}\frac{T_5}{T_1}\right)^{1-\gamma} = \left(\frac{1}{CR}T_R\frac{CR}{PR}\right)^{1-\gamma} = \left(\frac{T_R}{PR}\right)^{1-\gamma} \tag{5.34}$$

where Eq. (5.33) is also employed in the derivation of Eq. (5.34).

Now, substituting Eqs. (5.32)–(5.34) into Eq. (5.29), one obtains

$$\eta = 1 - \frac{PR \cdot CR^{1-\gamma} + (\gamma - 1)CR - \gamma\left(\frac{PR}{T_R}\right)}{PR - \left(\frac{PR}{T_R}\right)^{\gamma}} \tag{5.35}$$

5.3 Efficiency comparison

Table 5.2 summarizes the expressions obtained for the thermal efficiency of the gas power cycles in the preceding section. The third column in Table 5.2 gives PR as a function of CR, where appropriate. An efficiency comparison will be made assuming air as the working gas with $\gamma = 1.4$. The design constraints include (i) the amount of heat supplied is the same, and (ii) the maximum degree of volume change, CR, is identical in all engines.

Fig. 5.7 displays the efficiency of the cycles plotted against the normalized heat input defined as $q^* = q/(c_v T_1)$ at a fixed value of $CR = 21$. The Otto cycle exhibits a constant efficiency—see Eq. (5.14) and Table 5.2, and it possesses the highest thermal efficiency in Fig. 5.7 for any heat input $q^* < 4.12$. However, the Stirling cycle becomes the most efficient engine for $q^* > 4.12$.

Table 5.2 A summary of the thermal efficiency expressions and *PR-CR* relations.

Cycle	Efficiency	$PR = f(CR)$
Stirling	$1 - \frac{CR}{PR}$	$PR = CR \cdot T_R$
Brayton	$1 - \left(\frac{CR}{T_R}\right)^{\frac{1-\gamma}{2-\gamma}}$	$PR = (CR/T_R)^{\frac{\gamma}{2-\gamma}}$
Otto	$1 - CR^{1-\gamma}$	$PR = CR \cdot T_R$
Atkinson	$1 - \gamma \frac{T_R CR^{1-\gamma} - 1}{T_R - (T_R CR^{-\gamma})^{1-\gamma}}$	$PR = CR^{\gamma}$
Diesel	$1 - \frac{1}{\gamma}\left[\frac{T_R^{\gamma}(CR^{\gamma})^{1-\gamma} - 1}{T_R - CR^{\gamma-1}}\right]$	$PR = CR^{\gamma}$
Miller	$1 - \frac{PR.CR^{1-\gamma} + (\gamma-1)CR - \gamma\left(\frac{PR}{T_R}\right)}{PR - \left(\frac{PR}{T_R}\right)^{\gamma}}$	–

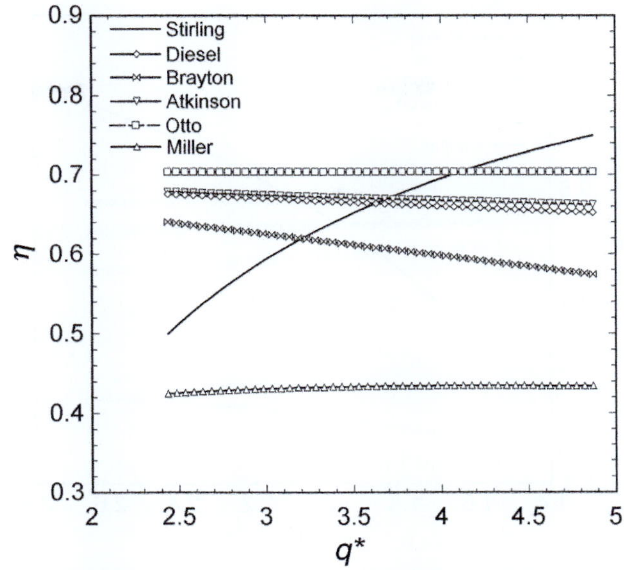

Fig. 5.7 Efficiency comparison of the power cycles undergoing identical degree of volume change ($CR=21$, $\gamma=1.4$).

A further observation in Fig. 5.7 is that the efficiency of the Atkinson and Diesel cycles is almost the same under the conditions of identical *CR* and heat input.

Yet, an alternative efficiency comparison can be made if the engines are constrained to operate between the same highest and lowest pressures. Fig. 5.8 compares the thermal efficiencies of the engines at a fixed pressure ratio of 21. The highest efficiency belongs to the Brayton cycle for any heat input $q^* < 2.08$, whereas the Stirling cycle is the most efficient engine for $q^* > 2.08$. Like in Fig. 5.7, the efficiency of the Atkinson and Diesel cycles is nearly identical and that the Miller cycle is the least efficient in Fig. 5.8.

It is important to realize that the answer to the question of *what engine is the most efficient* strictly depends on the design constraints. If the engines are constrained to operate between the same highest and lowest temperatures (identical T_R), the Stirling, Carnot and Ericsson cycles are the most efficient engines. If the engines are constrained to undergo the same degree of volume change (identical *CR*), the Otto cycle has the potential to possess the highest efficiency. Furthermore, the Brayton cycle may become the most efficient engine if the engines are subject to experience the same highest and lowest pressures.

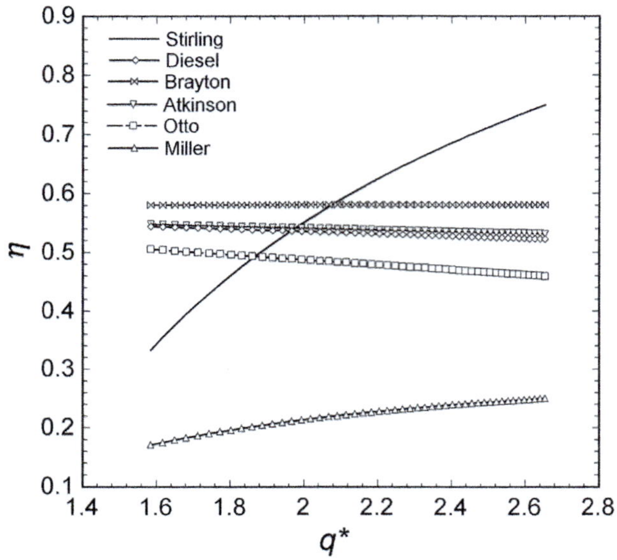

Fig. 5.8 Efficiency comparison of the power cycles undergoing identical degree of pressure change ($PR=21$, $\gamma=1.4$).

References

[1] S. Carnot, R.H. Thurston (Ed.), Reflections on the Motive Power of Heat, second ed., Wiley, New York, 1897.

[2] W.J.M. Rankine, Scientific Papers With a Memoir of the Author by P. G. Tait, Charles Griffin & Co., London, 1880.

[3] R.J. Allam, M.R. Palmer, G.W. Brown, J. Fetvedt, D. Freed, H. Nomoto, M. Itoh, N. Okita, C. Jones, High efficiency and low cost of electricity generation from fossil fuels while eliminating atmospheric emissions, including carbon dioxide, Energy Procedia 37 (2013) 1135–1149.

[4] A.I. Kalina, Low temperature geothermal system, US Patent 6,820,421.

[5] C. Borgnakke, R.E. Sonntag, Fundamentals of Thermodynamics, eighth ed., Wiley, New York, 2012.

CHAPTER SIX

Endoreversible heat engines

6.1 Introduction

The statement of the second law, unlike many physical laws, is not an explicit equality. It is based on the observation that while heat can flow spontaneously from a hot body to a clod body, it cannot flow back from the cold to the hot body without external expenditure. Also, while work can be completely converted into heat, it is impossible to convert all the heat back into work. The second law is a general observation that the thermal processes may proceed in certain directions. Nevertheless, it does not provide a quantitative measure on what amount of a given quantity of heat is convertible into work, or conversely, how much external work would be needed to transfer a unit of heat from a cold reservoir to a hot reservoir. As discussed in Chapter 2, it was Sadi Carnot [1] who first investigated maximum theoretical work that could be extracted from a given amount of heat. This investigation led him to come up with the design of an engine that would operate on a cycle comprising two adiabatic and two isothermal processes.

The significance of the Carnot cycle is that it played a key role in the formulation of the second law and the invention of entropy. Clausius [2] remarkable conclusion was that a Carnot-like (reversible) engine communicating with any number of heat reservoirs would yield no entropy production—see the discussion of Section 3.3.2. Clausius also concluded that the real heat engines would result in *uncompensated transformation*, or entropy generation.

It is natural to question whether there is any relation between the entropy produced by a heat engine and its thermal efficiency or power output. In 1975, Leff and Jones [3] discussed by means of an analytical argument that an increase in the thermal efficiency of an irreversible heat engine would not necessarily result in a decrease in its entropy production. Salamon et al. [4] showed that the maximum work and the minimum entropy production in heat engines might become equivalent under certain design conditions.

Entropy Analysis in Thermal Engineering Systems
https://doi.org/10.1016/B978-0-12-819168-2.00006-4

Numerous articles have appeared in the literature claiming a direct relationship between the entropy produced by a power-producing system and its thermal efficiency. In some cases, a relation between the maximum work production and the minimum entropy generation has been observed [5–8].

The present chapter investigates the equivalence of maximum thermal efficiency and minimum entropy production in typical endoreversible engines, including the models of Curzon-Ahlborn [9], Novikov [10], modified Novikov, and Carnot vapor cycle. It is aimed to show that the thermal efficiency of a heat engine may correlate with the entropy generation associated with the operation of that engine if the engine is endoreversible. The thermal efficiency and the power output of these engines at the condition of minimum entropy generation rate will also be examined.

6.2 Curzon-Ahlborn engine

The Curzon-Ahlborn engine [9] is depicted in Fig. 6.1 on a temperature-specific entropy (*T-s*) diagram. It is a Carnot engine, which experiences external irreversibility due to finite heat exchange between the engine and the hot and cold thermal reservoirs. Entropy is produced due to the finite time heat exchange between the endoreversible engine

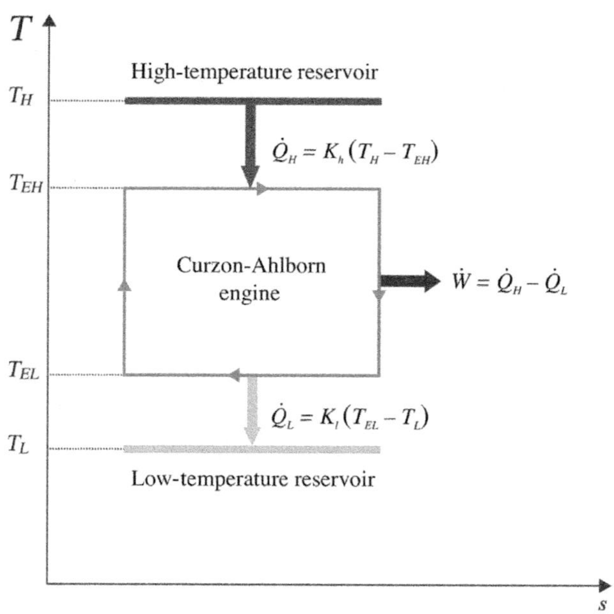

Fig. 6.1 The heat engine model of Curzon-Ahlborn on a *T-s* diagram.

(the rectangle region in Fig. 6.1) and the thermal reservoirs. Hence, the rate of entropy generation is

$$\dot{\Phi} = \left(\frac{\dot{Q}_H}{T_{EH}} - \frac{\dot{Q}_H}{T_H}\right) + \left(\frac{\dot{Q}_L}{T_L} - \frac{\dot{Q}_L}{T_{EL}}\right) \tag{6.1}$$

where $\dot{Q}_H$ is the heat rate received by the engine from the high-temperature thermal reservoir, and $\dot{Q}_L$ denotes the rate of heat rejected by the engine to the low-temperature thermal reservoir.

$$\dot{Q}_H = K_h(T_H - T_{EH}) \tag{6.2}$$

$$\dot{Q}_L = K_l(T_{EL} - T_L) \tag{6.3}$$

where K_h and K_l denote thermal conductance (assumed to be constant) at the hot-end and the cold-end sides of the engine, respectively. Also, T_{EH} and T_{EL} denote the highest and the lowest temperatures of the engine.

As the engine is endoreversible, we have

$$\frac{T_{EL}}{T_{EH}} = \frac{\dot{Q}_L}{\dot{Q}_H} = \frac{K_l(T_{EL} - T_L)}{K_h(T_H - T_{EH})} \tag{6.4}$$

Solving Eq. (6.4) for T_{EL} gives

$$T_{EL} = \frac{K_l T_L T_{EH}}{(K_h + K_l)T_{EH} - K_h T_H} \tag{6.5}$$

Using Eq. (6.4), Eq. (6.1) reduces to

$$\dot{\Phi} = \frac{\dot{Q}_L}{T_L} - \frac{\dot{Q}_H}{T_H} \tag{6.6}$$

A combination of Eqs. (6.2), (6.3), (6.5), and (6.6) allows expressing Eq. (6.6) as

$$\dot{\Phi} = K_h(T_H - T_{EH}) \left[\frac{1}{\left(1 + \frac{K_h}{K_l}\right)T_{EH} - \frac{K_h}{K_l}T_H} - \frac{1}{T_H}\right] \tag{6.7}$$

Solving $\partial\dot{\Phi}/\partial T_{EH} = 0$ leads to $(T_{EH})_{opt} = T_H$. Substituting this result into Eq. (6.5), we also find $(T_{EL})_{opt} = T_L$. Minimization of the entropy generation rate associated with the Curzon-Ahlborn model suggests that any irreversibility between the thermal reservoirs and the engine should be removed, which will consequently lead to zero finite time power. On the other hand, the power produced by the engine is $\dot{W} = \dot{Q}_H - \dot{Q}_L$. Using Eqs. (6.2), (6.3), and (6.5), it can be shown that

$$\dot{W} = K_h(T_H - T_{EH}) - K_l\left[\frac{K_l T_L T_{EH}}{(K_h + K_l)T_{EH} - K_h T_H} - T_L\right] \tag{6.8}$$

Applying $\partial\dot{W}/\partial T_{EH} = 0$ yields

$$(T_{EH})_{opt} = \frac{K_l\sqrt{T_L T_H} + K_h T_H}{K_h + K_l} \tag{6.9}$$

Substituting Eq. (6.9) into Eq. (6.8) leads to an expression for the maximum power production.

$$\dot{W}_{max} = \frac{K_h K_l}{K_h + K_l}\left(\sqrt{T_H} - \sqrt{T_L}\right)^2 \tag{6.10}$$

Curzon and Ahlborn [9] showed that the engine efficiency at maximum power is $(\eta_{CA})_{\dot{W}_{max}} = 1 - \sqrt{T_L/T_H}$, where the subscript CA refers to the Curzon-Ahlborn engine (please refer to Appendix A for a discussion on the efficiency at maximum power). To find out whether there is any relationship between the entropy production of the Curzon-Ahlborn cycle and its thermal efficiency, we represent Eq. (6.7) in a dimensionless form by dividing it by K_l.

$$\Phi^* = r_K(r_T - T^*_{EH})\left[\frac{1}{(1 + r_K)T^*_{EH} - r_K r_T} - \frac{1}{r_T}\right] \tag{6.11}$$

where

$$\Phi^* = \frac{\dot{\Phi}}{K_l};\ r_K = \frac{K_h}{K_l};\ r_T = \frac{T_H}{T_L};\ T^*_{EH} = \frac{T_{EH}}{T_L}$$

Using Eq. (6.5), the thermal efficiency of the engine may be represented as

$$\eta_{CA} = 1 - \frac{1}{(1 + r_K)T^*_{EH} - r_K r_T} \tag{6.12}$$

We may also rewrite Eq. (6.8) in a normalized form as

$$W^* = 1 + r_K(r_T - T^*_{EH}) - \frac{T^*_{EH}}{(1 + r_K)T^*_{EH} - r_K r_T} \tag{6.13}$$

where $W^* = \frac{\dot{W}}{K_l T_L}$.

Fig. 6.2 shows the variation of the thermal efficiency, the normalized power output, and the normalized entropy production rate of the Curzon-Ahlborn engine with T^*_{EH} for typical values of $r_T = 6$ and $r_K = 6$. It is seen that the entropy production monotonically decreases, whereas the thermal efficiency increases by increasing T^*_{EH}. On the other hand, the power output peaks at $T^*_{EH} = 4.86$. Note that as $T^*_{EH} \rightarrow r_T$, the efficiency approaches the Carnot efficiency, and the rate of entropy produced by the engine approaches zero. However, at this condition, the power output of the engine reaches zero, too. From this analysis, we conclude that for the Curzon-Ahlborn engine, the minimization of the entropy production rate is equivalent to the maximization of the thermal efficiency, but not to the maximization of power output.

Note that the variable in our thermodynamic optimization is T^*_{EH}, and we assumed that r_K is a constant. One may treat T^*_{EH} as a fixed parameter and optimize the system by varying r_K. This is also a possible; however, our conclusion mentioned in the previous paragraph is still correct. In other words, even with varying r_K and for fixed values of T^*_{EH} and r_T, it can be shown with a similar analysis that the minimum entropy production correlates with only the maximum thermal efficiency, not with the maximum power output. This is graphically demonstrated in Fig. 6.3.

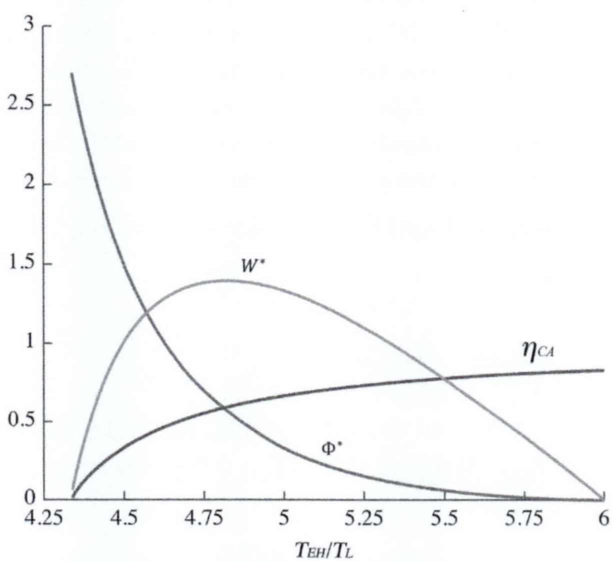

Fig. 6.2 Variation of the thermal efficiency, normalized power output, and normalized entropy production rate of the Curzon-Ahlborn engine with $T^*_{EH} = T_{EH}/T_L$ ($r_T = 6$, $r_K = 2$).

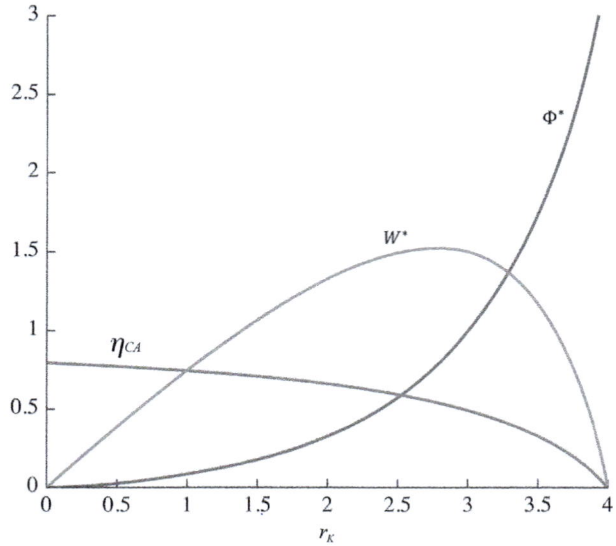

Fig. 6.3 Variation of the thermal efficiency, normalized power output, and normalized entropy production rate of the Curzon-Ahlborn engine with r_K ($r_T = 6$, $T^*_{EH} = 5$).

6.3 Novikov's engine

The operation of the Novikov's engine [10] is depicted on a T-s diagram in Fig. 6.4. In this model, the temperature at the cold-end side of the engine is the same as the low-temperature reservoir's temperature, T_L. In other words, it is a Carnot engine which experiences external irreversibility due to the finite heat exchange between the engine and the hot thermal reservoir. As the engine is internally reversible, its efficiency is $\eta_N = 1 - T_L/T_{EH}$, and the power produced by the engine is $\dot{W} = \dot{Q}_H\left(1 - \frac{T_L}{T_{EH}}\right)$. Using Eq. (6.2), we find

$$\dot{W} = K_h(T_H - T_{EH})\left(1 - \frac{T_L}{T_{EH}}\right) \tag{6.14}$$

Assuming a constant K_h and fixed T_H and T_L, Eq. (6.14) has an optimum with respect to T_{EH}. Applying $\partial\dot{W}/\partial T_{EH} = 0$ yields an equation whose solution gives

$$(T_{EH})_{opt} = \sqrt{T_L T_H} \tag{6.15}$$

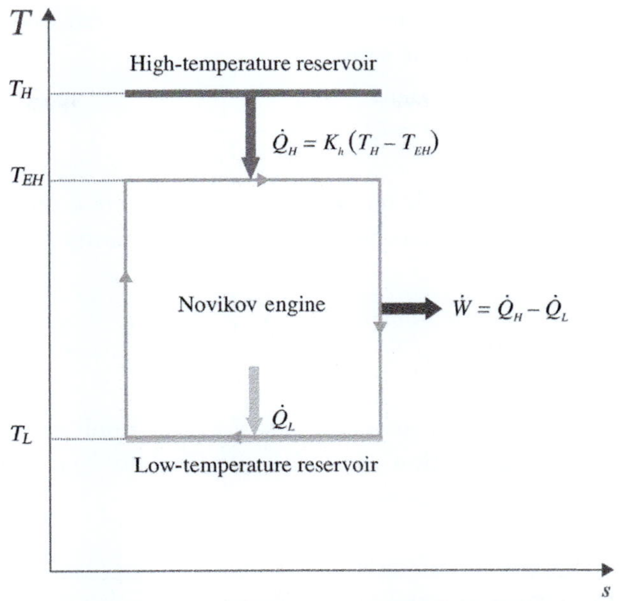

Fig. 6.4 The heat engine model of Novikov on a *T-s* diagram.

Eq. (6.15) allows us to find the efficiency and the power output of the Novikov's engine at maximum power condition.

$$(\eta_N)_{\dot{W}_{max}} = 1 - \sqrt{\frac{T_L}{T_H}} \tag{6.16}$$

$$\dot{W}_{max} = K_h\left(\sqrt{T_H} - \sqrt{T_L}\right)^2 \tag{6.17}$$

An interesting observation is that the efficiencies of the engine models shown in Figs. 6.1 and 6.4 at maximum power output are the same. However, comparing Eqs. (6.17) and (6.10), it can be inferred that the maximum power produced by the Novikov's engine is greater than that of the Curzon-Ahlborn engine. This is because at the condition of maximum power, the highest temperature of the engine, T_{EH}, of the Curzon-Ahlborn engine is higher than that of the Novikov's engine; compare Eqs. (6.9) and (6.15), meaning that the heat input of the former is less than that of the latter. As the efficiency of both engines at maximum power is the same, it can be implied that the maximum power of the Curzon-Ahlborn engine is less than that of the Novikov's engine.

In the next step, we calculate the entropy generation rate associated with the operation of the Novikov's model.

$$\dot{\Phi} = \dot{Q}_H \left(\frac{1}{T_{EH}} - \frac{1}{T_H} \right) \tag{6.18}$$

Notice that the cold-end side temperature of the engine is the same as the low-temperature thermal reservoir's temperature. Substituting Eq. (6.2) into Eq. (6.18) yields

$$\dot{\Phi} = K_h (T_H - T_{EH}) \left(\frac{1}{T_{EH}} - \frac{1}{T_H} \right) \tag{6.19}$$

Solving $\partial\dot{\Phi}/\partial T_{EH} = 0$ leads to $(T_{EH})_{opt} = T_H$. This result reveals that the minimum entropy generation rate takes place when $T_{EH} \rightarrow T_H$, which would give a zero finite time power.

6.4 Modified Novikov's engine

Let us now consider a modified model of Novikov's engine, in which the finite time heat exchange only takes place at the cold-end side of the engine (see Fig. 6.5). Thus, the efficiency and the power output of the

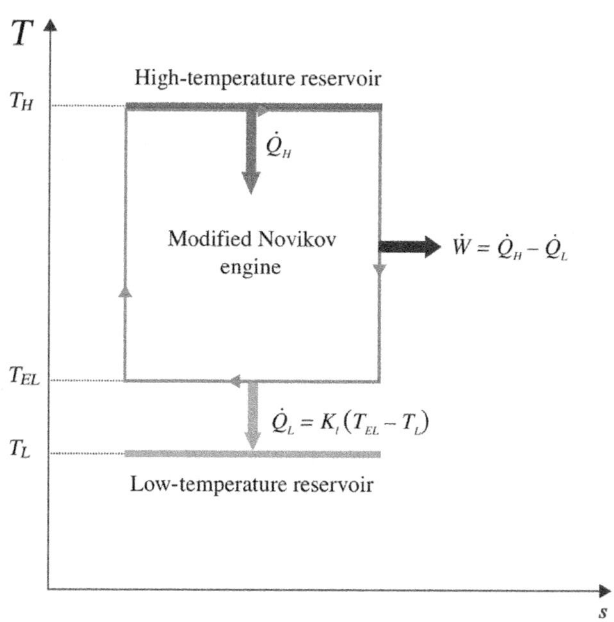

Fig. 6.5 Modified Novikov's engine on a *T-s* diagram.

engine are given by $\eta_{MN}=1-T_{EL}/T_H$ and $\dot{W}=\dot{Q}_H\left(1-\frac{T_{EL}}{T_H}\right)$, respectively. Noting that $\dot{Q}_H=\dot{W}+\dot{Q}_L$ and using Eq. (6.3), we find

$$\dot{W}=K_l(T_H-T_{EL})\left(1-\frac{T_L}{T_{EL}}\right) \tag{6.20}$$

Maximization of the power output given in Eq. (6.20) with respect to T_{EL} yields $(T_{EL})_{opt}=\sqrt{T_L T_H}$, which is the same as $(T_{EH})_{opt}$ of the Novikov's engine. The efficiency and the power output of the modified Novikov's engine at maximum power production are obtained by

$$(\eta_{MN})_{\dot{W}_{max}}=1-\sqrt{\frac{T_L}{T_H}} \tag{6.21}$$

$$\dot{W}_{max}=K_l\left(\sqrt{T_H}-\sqrt{T_L}\right)^2 \tag{6.22}$$

Comparing Eqs. (6.22) and (6.10), it can be inferred that when the thermal conductance at the hot-end side of the Curzon-Ahlborn model tends to infinity $K_h\rightarrow\infty$, the model of Curzon-Ahlborn reduces to the modified model of Novikov's engine. A further observation is that the efficiency of all three engines that we have examined so far at maximum power output is $1-\sqrt{T_L/T_H}$, whereas the maximum power of the Novikov's engine is the highest, and that of the modified Novikov's engine is the lowest, and that of the Curzon-Ahlborn engine in between.

To evaluate the production rate of entropy, we note that there is merely one source of entropy generation due to the finite time heat exchange at the cold-end side of the engine. The heat transfer at the hot-end side of the engine takes place reversibly. So, $\dot{\Phi}=\dot{Q}_L(1/T_L-1/T_{EL})$, which using Eq. (6.3), we get

$$\dot{\Phi}=K_l(T_{EL}-T_L)\left(1-\frac{T_L}{T_{EL}}\right) \tag{6.23}$$

Solving $\partial\dot{\Phi}/\partial T_{EL}=0$ leads to $(T_{EL})_{opt}=T_L$ at which the power production is zero; see Eq. (6.20). It can be concluded that the minimum entropy generation and the maximum power output are two different operational regimes in the modified model of Novikov's engine.

Like the Curzon-Ahlborn model, the relationship between the entropy production and the thermal efficiency is graphically demonstrated in Fig. 6.6 for the Novikov's engine and the modified Novikov's engine. For the case of the Novikov's engine (Fig. 6.6A), the normalized power output W^* and

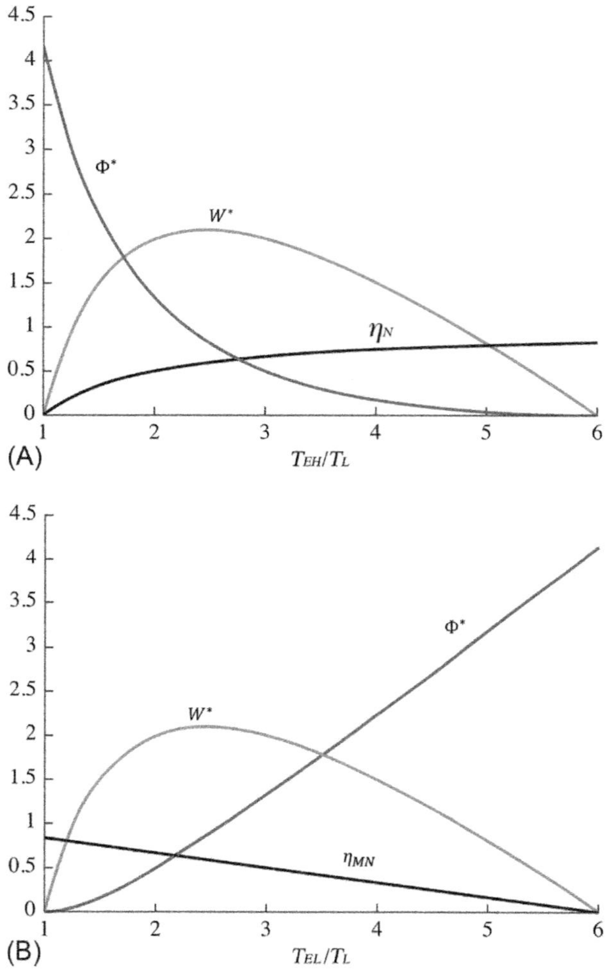

Fig. 6.6 Variation of the thermal efficiency, normalized power output, and normalized entropy production rate at $r_T = 6$ (A) Novikov' engine, and (B) modified Novikov's engine.

the normalized entropy production rate Φ^* are defined the same way as before. However, because $K_l \rightarrow \infty$ for the case of the modified Novikov's engine (Fig. 6.6B), these two parameters are defined as $W^* = \dot{W}/K_h T_L$ and $\Phi^* = \Phi/K_h$. Also, note that the results for the Novikov's model are presented as a function of T_{EH}/T_L, whereas those for the modified Novikov's model are depicted versus T_{EL}/T_L.

Fig. 6.6 reveals that an increase in the entropy production is equivalent to a decrease in the thermal efficiency in both Novikov's and modified

Novikov's engines. For the Novikov's engine, the thermal efficiency monotonically increases, whereas the entropy production consistently decreases with T_{EH}/T_L. The power output of the engine peaks at $T_{EH}/T_L=2.1$. For the modified Novikov's engine, with an increase in T_{EL}/T_L, the thermal efficiency decreases whereas the entropy production increases. The power output attains its maximum at $T_{EL}/T_L=2.1$. From this discussion, we arrive at the conclusion that for the engine models of Novikov and modified Novikov, the minimization of the entropy production is equivalent to the maximization of the thermal efficiency, a similar conclusion that we reached in Section 6.2 for the Curzon-Ahlborn engine.

6.5 Carnot vapor cycle

The operation of the Carnot vapor cycle is shown on a *T-s* diagram in Fig. 6.7. Unlike the heat engine models examined in previous sections, which interact with fixed temperature reservoirs, the model of Fig. 6.7 exchanges heat with a hot fluid and a cold fluid whose specific heat is denoted by $c_{p,h}$ and $c_{p,l}$, respectively. The cycle undergoes an isothermal evaporation process at temperature T_{EH}, and an isothermal condensation process

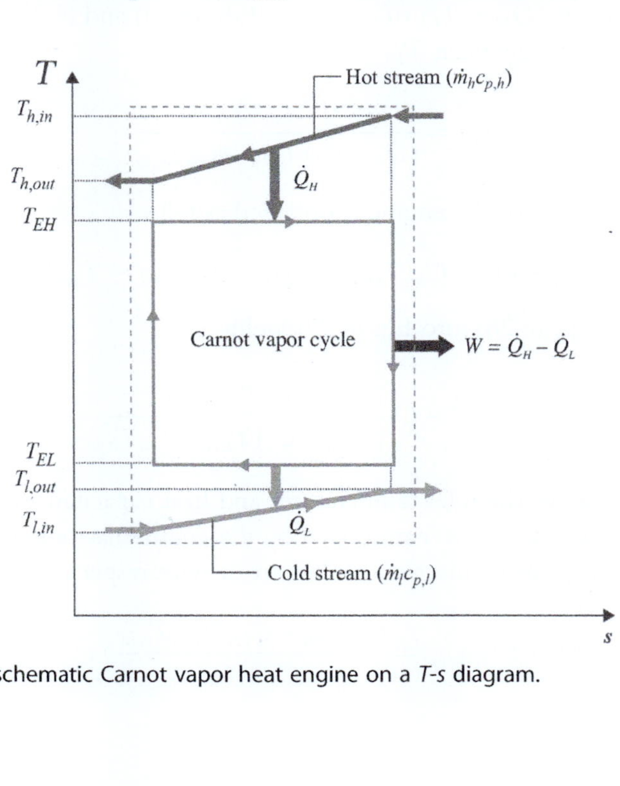

Fig. 6.7 A schematic Carnot vapor heat engine on a *T-s* diagram.

at temperature T_{EL}. The evaporation process takes place by receiving heat from the hot stream through a heat exchanger. The inlet and outlet temperatures of the hot stream are $T_{h,in}$ and $T_{h,out}$, respectively. At the cold-end side of the engine, the condensation heat is rejected through another heat exchanger to the cold stream, which enters the heat exchanger at temperature $T_{l,in}$ and exits at temperature $T_{l,out}$.

Assume that both heat exchangers at the hot-end and the cold-end sides are operating ideally. In other words, the temperature of the hot stream leaving the engine is equal to the evaporation temperature, i.e., $T_{h,out} = T_{EH}$, and the exit temperature of the cold stream is the same as the condensation temperature, i.e., $T_{l,out} = T_{EL}$. In this case, the heat rate supplied from the hot stream for evaporation of the working fluid is

$$\dot{Q}_H = \dot{C}_h(T_{H,in} - T_{EH}) \tag{6.24}$$

where $\dot{C} = \dot{m}c_p$. The rate of heat rejected by the engine at the cold-end side heat exchanger is absorbed by the cold stream. Hence,

$$\dot{Q}_L = \dot{C}_l(T_{EL} - T_{l,in}) \tag{6.25}$$

The power-producing compartment (the rectangle in Fig. 6.7) is internally reversible, so $\dot{Q}_L/\dot{Q}_H = T_{EL}/T_{EH}$. Using Eqs. (6.24) and (6.25), a relation can be established between T_{EL} and T_{EH}.

$$T_{EL} = \frac{\dot{C}_l T_{l,in} T_{EH}}{\left(\dot{C}_h + \dot{C}_l\right) T_{EH} - \dot{C}_h T_{h,in}} \tag{6.26}$$

The power output of the engine is obtained as follows.

$$\dot{W} = \dot{Q}_H - \dot{Q}_L = \dot{C}_h(T_{H,in} - T_{EH}) - \dot{C}_l(T_{EL} - T_{l,in}) \tag{6.27}$$

Substituting Eqs. (6.26) into Eq. (6.27) yields

$$\dot{W} = \dot{C}_h(T_{H,in} - T_{EH}) - \dot{C}_l\left[\frac{\dot{C}_l T_{l,in} T_{EH}}{\left(\dot{C}_h + \dot{C}_l\right) T_{EH} - \dot{C}_h T_{h,in}} - T_{l,in}\right] \tag{6.28}$$

For fixed values of the inlet temperatures and heat capacitances of the hot and cold streams, the power produced by the engine has only one degree of freedom, T_{EH}. Maximization of the power with respect to T_{EH} yields

$$\left(T_{EH}\right)_{opt} = \frac{\dot{C}_h T_{h,in} + \dot{C}_l \sqrt{T_{l,in} T_{h,in}}}{\dot{C}_h + \dot{C}_l} \tag{6.29}$$

The maximum power output of the cycle is obtained by substituting Eq. (6.29) into Eq. (6.28).

$$\dot{W}_{max} = \frac{\dot{C}_h \dot{C}_l}{\dot{C}_h + \dot{C}_l} \left(\sqrt{T_{h,in}} - \sqrt{T_{l,in}}\right)^2 \tag{6.30}$$

An important observation in Fig. 6.7 is that unlike the engine models discussed in Sections 6.2–6.4 where all the heat supplied from the thermal reservoir is transferred to the engine, here only a fraction of the heat is transferred to the power producing compartment. If the hot stream is supplied from the ambient (e.g., air) as in most steam engines, it is first heated (for instance, in the furnace of a steam power plant) to a desired temperature $T_{h,in}$, and eventually it is exhausted to the atmosphere that is at temperature $T_{l,in}$ (see Fig. 6.8). So, the thermal efficiency of the entire plant is given by $\eta_{th} = \dot{W}/\dot{Q}_{in}$, where $\dot{Q}_{in}$ denotes the rate of heat transfer from the high-temperature reservoir to the hot stream. Hence,

$$\dot{Q}_{in} = \dot{Q}_H + \dot{C}_h(T_{EH} - T_{l,in}) = \dot{C}_h(T_{h,in} - T_{l,in}) \tag{6.31}$$

Using Eqs. (6.30) and (6.31), we find an expression for the thermal efficiency of the entire plant at maximum power as follows.

$$(\eta_{th})_{\dot{W}_{max}} = \frac{\dot{C}_l}{\dot{C}_h + \dot{C}_l} \left(\frac{\sqrt{T_{h,in}} - \sqrt{T_{l,in}}}{\sqrt{T_{h,in}} + \sqrt{T_{l,in}}} \right) \tag{6.32}$$

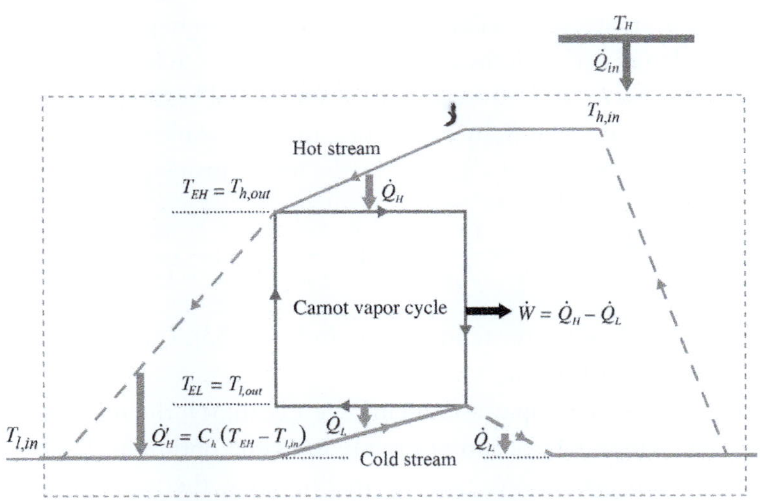

Fig. 6.8 Illustration of different sources of entropy generation associated with the operation of the Carnot vapor cycle depicted in Fig. 6.7.

Note that the efficiency of the power-producing compartment at maximum power is $1-\sqrt{T_{l,in}/T_{h,in}}$. Also, as the heat input $\dot{Q}_{in}$ is constant, the maximum efficiency occurs at the same optimum T_{EH} that the power output is maximized, i.e., Eq. (6.29).

The entropy production rate due to the heat exchange between the power-producing compartment and the hot and cold streams is $\dot{\Phi}=\Delta\dot{S}_h+\Delta\dot{S}_l$, where $\Delta\dot{S}_h$ and $\Delta\dot{S}_l$ denote the net change in the entropies of the hot and cold streams, respectively, due to the heat exchange with the working fluid of the cycle.

To evaluate the total entropy generation rate associated with the operation of the engine model of Fig. 6.7 and based on the arguments of Bejan [8], we need to account for additional sources of entropy generation due to (i) the transfer of heat from the exhaust of the hot stream to the surrounding, (ii) the rejection of heat from the exhaust of the cold stream to the surrounding, and (iii) heating the hot stream to increase its temperature from $T_{l,in}$ to $T_{h,in}$; see Fig. 6.8.

As the hot stream is provided from the ambient, its temperature first rises from $T_{l,in}$ to $T_{h,in}$. It then reduces to $T_{h,out}$ within the hot-end side heat exchanger where it loses part of its energy to evaporate the working substance of the engine. Finally, it cools down to $T_{l,in}$ after it returns back to the ambient. Thus, the net change in the entropy of the hot stream fluid is zero. Likewise, the net change in the entropy of the cold stream is zero as it is supplied from the ambient and eventually discharged to the ambient. So, a more accurate way to determine the total entropy generation rate of the system designated with the dashed rectangle in Fig. 6.8 is to evaluate the total heat transfer to the hot stream as well as the total heat rejected from the cycle to the ambient. Thus, by accounting for all possible sources related to the operation of the Carnot vapor cycle, the total entropy production rate is

$$\dot{\Phi}_{tot}=\frac{\dot{Q}_{out}}{T_{l,in}}-\frac{\dot{Q}_{in}}{T_H}=\dot{Q}_{in}\left(\frac{1}{T_{l,in}}-\frac{1}{T_H}\right)-\frac{\dot{W}}{T_{l,in}} \tag{6.33}$$

where T_H is the heat source temperature. In Eq. (6.33), $\dot{Q}_{out}$ was eliminated using $\dot{W}=\dot{Q}_{in}-\dot{Q}_{out}$.

As the rate of heat input is constant; see Eq. (6.31), the minimum total entropy generation rate occurs at exactly the same optimum T_{EH} that the power output is maximized, i.e., Eq. (6.29). Thus, the maximum power, maximum thermal efficiency, and minimum entropy production rate become coincident for the Carnot vapor cycle when the heat input is constant. In the case of constant heat input, optimization of any model of engine that interacts with two thermal sources, based on maximum thermal efficiency, maximum power output and minimum entropy production would result in an identical design [5].

From practical viewpoint, optimization of an engine with a fixed heat input is irrelevant. A given heat input is equivalent to a fixed mass of fuel to be burnt in furnace. Once the problem is reduced to a fixed amount of burning fuel, an important ability of varying mass of fuel at other operational conditions (such as at part-load operation) is taken away. So, a, more general case would be to treat this parameter as a variable together with many other parameters, which may directly or indirectly influence the thermal efficiency of an engine.

Let us now consider a Carnot vapor cycle with varying heat input. For this, we assume that $\dot{C}_h$ is the design parameter, and all other parameters, including T_{EH} are constant. Similar to the case with varying T_{EH}, it can be inferred from Eq. (6.28) that there is an optimum $\dot{C}_h$ given by Eq. (6.34), which maximizes the power output.

$$\left(\dot{C}_h\right)_{opt} = \dot{C}_l\left(\frac{T_{EH} - \sqrt{T_{EH}T_{l,in}}}{T_{h,in} - T_{EH}}\right) \tag{6.34}$$

Substituting Eq. (6.34) into Eq. (6.28), the maximum power output at optimum heat capacity of the hot stream is obtained as

$$\dot{W}_{max} = \dot{C}_l\left(\sqrt{T_{EH}} - \sqrt{T_{l,in}}\right)^2 \tag{6.35}$$

Eq. (6.35) reveals that when the power optimization is performed with $\dot{C}_h$, the maximum power depends only on three parameters: the heat capacity and the inlet temperature of the cold stream, and the highest temperature of the engine T_{EH}. On the other hand, the thermal efficiency of the cycle at maximum power is

$$(\eta_{th})_{\dot{W}_{max}} = \left(1 - \sqrt{\frac{T_{l,in}}{T_{EH}}}\right)\left(\frac{T_{h,in} - T_{EH}}{T_{h,in} - T_{l,in}}\right) \tag{6.36}$$

It can be implied from Eq. (6.36) that the thermal efficiency at maximum power is independent of the heat capacities of the cold and hot streams, but it depends on their inlet temperatures as well as the highest temperature of the engine.

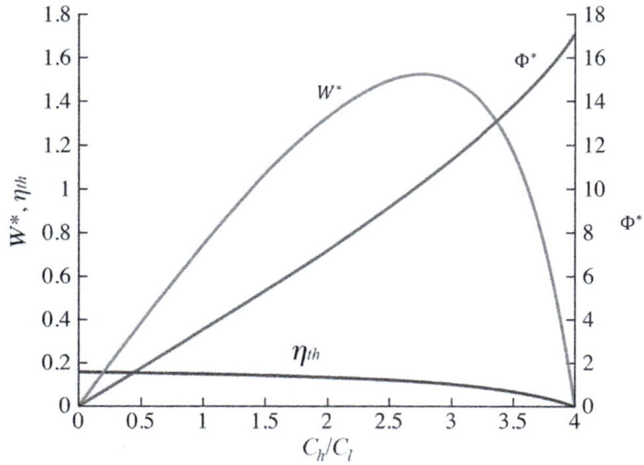

Fig. 6.9 Variation of the thermal efficiency, normalized power output, and normalized entropy production rate of the Carnot vapor cycle with $\dot{C}_h/\dot{C}_l$. $\dot{\Phi}^*$ is the total entropy production given by Eq. (6.33) divided by $\dot{C}_l$ ($T_{EH}/T_{l,in}=5$, $T_{h,in}/T_{l,in}=6$, $T_H/T_{l,in}=7$).

To illustrate the dependence of efficiency, power output, and the entropy generation rate of the Carnot vapor cycle with our design parameter, $\dot{C}_h$, it is convenient to define the normalized power output and the normalized entropy generation rate $W^* = \dot{W}/\left(\dot{C}_l T_{l,in}\right)$ and $\Phi^* = \dot{\Phi}_{tot}/\dot{C}_l$, where $\dot{W}$ and $\dot{\Phi}_{tot}$ are given in Eqs. (6.28) and (6.33), respectively.

The variation of the thermal efficiency, normalized power output, and normalized entropy production rate of the Carnot vapor cycle versus the ratio $\dot{C}_h/\dot{C}_l$ is illustrated in Fig. 6.9 for typical values of $T_{EH}/T_{l,in}=5$, $T_{h,in}/T_{l,in}=6$, and $T_H/T_{l,in}=7$. The results in Fig. 6.9 demonstrate that the thermal efficiency inversely correlates with Φ^*. The thermal efficiency of the engine decreases monotonically with $\dot{C}_h/\dot{C}_l$, whereas Φ^* consistently increases with $\dot{C}_h/\dot{C}_l$. On the other hand, the power produced by the cycle peaks at $\dot{C}_h/\dot{C}_l=2.8$. Thus, when the heat input is a varying parameter, the regimes of maximum thermal efficiency and minimum entropy generation rate are equivalent, but they are different from the regime of maximum power.

References

[1] S. Carnot, Reflections on the Motive Power of Heat. R. H. Thurston, second ed., Wiley, New York, 1897.

[2] R. Clausius, The Mechanical Theory of Heat, Translated by W. R. Brown, MacMillan & Co., London, 1879.

[3] H.S. Leff, G.L. Jones, Irreversibility, entropy production, and thermal efficiency, Am. J. Phys. 43 (1975) 973–980.
[4] P. Salamon, K.H. Hoffmann, S. Schubert, R.S. Berry, B. Andresen, What conditions make minimum entropy production equivalent to maximum power production, J. Non-Equil. Thermodyn. 26 (2001) 73–83.
[5] Y. Haseli, Performance of irreversible heat engines at minimum entropy generation, Appl. Math. Model. 37 (2013) 9810–9817.
[6] Y. Haseli, Efficiency of irreversible Brayton cycles at minimum entropy generation, Appl. Math. Model. 40 (2016) 8366–8376.
[7] X.B. Liu, J.A. Meng, Z.Y. Guo, Entropy generation extremum and entransy dissipation extremum for heat exchanger optimization, Chinese Sci. Bull. 54 (2009) 943–947.
[8] A. Bejan, The equivalence of maximum power and minimum entropy generation rate in the optimization of engines, J. Energy Resour. Technol. 118 (1996) 98–101.
[9] F.L. Curzon, B. Ahlborn, Efficiency of a Carnot engine at maximum power output, Am. J. Phys. 43 (1975) 22–24.
[10] I.I. Novikov, The efficiency of atomic power stations, J. Nucl. Energy 7 (1958) 125–128.

CHAPTER SEVEN

Irreversible engines—Closed cycles

7.1 Introduction

The endoreversible heat engines discussed in Chapter 6 were assumed to experience external irreversibility only. The present chapter extends the analysis of the preceding chapter to heat engines which undergo both external and internal irreversibilities. The analysis will specifically focus on four widely known gas cycles, including Brayton, Otto, Atkinson, and Diesel engines operating in a closed cycle while exchanging heat with two thermal reservoirs. Expressions were derived in Chapter 5 for the efficiencies of the ideal design of these cycles. The performance of each engine at maximum thermal efficiency, maximum power output, and minimum entropy generation rate will individually be investigated.

The underlying assumptions to be adopted for simplicity of the analysis include (i) air is assumed to be the working gas with constant properties, (ii) the air behaves like an ideal gas, (iii) pressure drop is negligible, (iv) external irreversibilities are due to heat transfer processes between the engine and the high- and low-temperature thermal reservoirs characterized by T_H and T_L, (v) internal irreversibilities are due to the compression and expansion processes, (vi) the temperatures of the thermal reservoirs are fixed. As depicted in Fig. 7.1, all four cycles have a similar T-s diagram.

The irreversible compression and expansion processes take place through lines $1 \rightarrow 2$ and $3 \rightarrow 4$, respectively. The dotted lines $1 \rightarrow 2\text{s}$ and $3 \rightarrow 4\text{s}$ show the isentropic compression and expansion processes. Lines $2 \rightarrow 3$ and $4 \rightarrow 1$ represent the heat addition and heat removal processes, respectively. In the Brayton cycle, the heat transfer processes are isobaric whereas they are isochoric in the Otto cycle. In the Atkinson cycle, heat is added to the engine at constant volume (isochoric), and heat is removed at constant

Entropy Analysis in Thermal Engineering Systems
https://doi.org/10.1016/B978-0-12-819168-2.00007-6

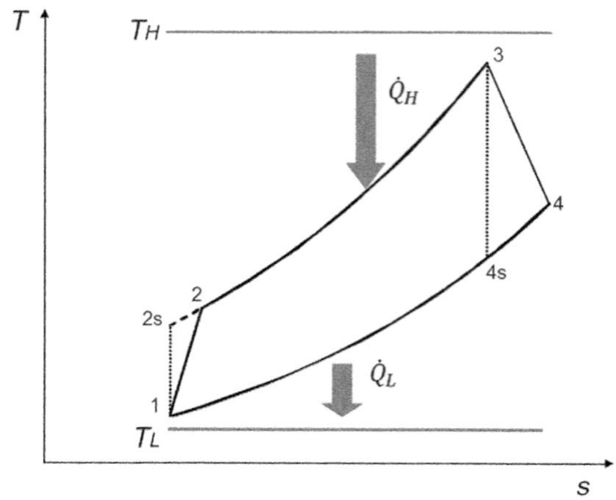

Fig. 7.1 A schematic of the *T-s* diagram of the irreversible closed Brayton, Otto, Atkinson, and Diesel cycles operating between a high- and a low-temperature reservoir.

Table 7.1 The processes of the irreversible Brayton, Otto, Atkinson, and Diesel cycles.

	Processes			
Engine	**Compression**	**Heat addition**	**Expansion**	**Heat removal**
Brayton	Nonisentropic	Isobaric	Nonisentropic	Isobaric
Otto	Nonisentropic	Isochoric	Nonisentropic	Isochoric
Atkinson	Nonisentropic	Isochoric	Nonisentropic	Isobaric
Diesel	Nonisentropic	Isobaric	Nonisentropic	Isochoric

pressure (isobaric). In the Diesel cycle, the heat addition is isobaric, whereas the heat removal is isochoric. Table 7.1 summarizes the processes of the four engines.

7.2 Brayton cycle

The isentropic efficiencies of the compression and expansion processes are defined as

$$\eta_{com} = \frac{T_{2s} - T_1}{T_2 - T_1} \tag{7.1}$$

$$\eta_{exp} = \frac{T_3 - T_4}{T_3 - T_{4s}} \tag{7.2}$$

The temperature of the air at states 2 and 4 can be obtained using the following relationships.

$$T_2 = T_1\left(1 + \frac{PR^{\frac{\gamma-1}{\gamma}} - 1}{\eta_{com}}\right) \tag{7.3}$$

$$T_4 = T_3\left[1 - \eta_{exp}\left(1 - PR^{\frac{1-\gamma}{\gamma}}\right)\right] \tag{7.4}$$

where $PR = p_2/p_1$ denotes the pressure ratio.

The rate of heat received from the external heat reservoir is determined as follows.

$$\dot{Q}_H = \dot{m}c_p(T_3 - T_2) = \dot{m}c_p T_1\left(T_R - 1 - \frac{PR^{\frac{\gamma-1}{\gamma}} - 1}{\eta_{com}}\right) \tag{7.5}$$

where $\dot{m}$ is the mass flowrate of the air, c_p denotes the specific heat at constant-pressure, and $T_R = T_3/T_1$ is the temperature ratio. Note also that Eq. (7.3) is used in Eq. (7.5).

The rate of heat rejected by the cycle to the low-temperature reservoir is obtained as

$$\dot{Q}_L = \dot{m}c_p(T_4 - T_1) = \dot{m}c_p T_1\left[T_R - 1 - \eta_{exp}T_R\left(1 - PR^{\frac{1-\gamma}{\gamma}}\right)\right] \tag{7.6}$$

The net power produced by the cycle and its thermal efficiency are determined as follows.

$$\dot{W}_{net} = \dot{Q}_H - \dot{Q}_L = \dot{m}c_p T_1\left(1 - PR^{\frac{1-\gamma}{\gamma}}\right)\left(\eta_{exp}T_R - \frac{PR^{\frac{\gamma-1}{\gamma}}}{\eta_{com}}\right) \tag{7.7}$$

$$\eta = \frac{\dot{W}_{net}}{\dot{Q}_H} = \frac{\left(1 - PR^{\frac{1-\gamma}{\gamma}}\right)\left(\eta_{exp}T_R - \frac{PR^{\frac{\gamma-1}{\gamma}}}{\eta_{com}}\right)}{T_R - 1 - \frac{PR^{\frac{\gamma-1}{\gamma}} - 1}{\eta_{com}}} \tag{7.8}$$

It can be inferred from Eq. (7.7) that the power output of the engine is zero at $PR = 1$ and $PR = \left(\eta_{com}\eta_{exp}T_R\right)^{\frac{\gamma}{\gamma-1}}$. Thus, there exists an extremum for the power output between these two pressure ratios. This is also graphically

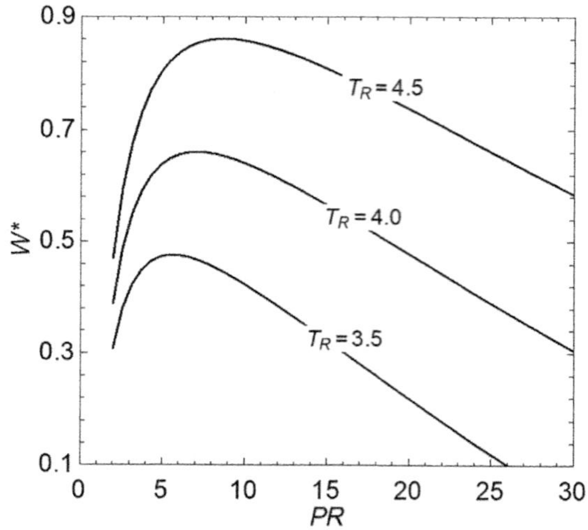

Fig. 7.2 Variation of the normalized power output, $W^* = \dot{W}_{net}/\dot{m}c_pT_1$, of the irreversible Brayton cycle with the pressure ratio, $\gamma = 1.4$, $\eta_{com} = 0.85$, and $\eta_{exp} = 0.90$.

demonstrated in Fig. 7.2. The optimum pressure ratio that yields a maximum power can be found by applying $\partial \dot{W}_{net}/\partial PR = 0$. Hence,

$$(PR)_{\dot{W}_{max}} = \left(\eta_{com}\eta_{exp}T_R\right)^{\frac{\gamma}{2(\gamma-1)}} \tag{7.9}$$

Substituting Eq. (7.9) into Eq. (7.7) leads to an expression for the maximum power output.

$$\dot{W}_{max} = \dot{m}c_pT_1\eta_{exp}T_R\left(1 - \frac{1}{\sqrt{\eta_{com}\eta_{exp}T_R}}\right)^2 \tag{7.10}$$

For instance, substitute $\gamma = 1.4$, $\eta_{com} = 0.85$ and $\eta_{exp} = 0.90$ into Eq. (7.9). The optimum pressure ratio at a temperature ratio of 4 is

$$(PR)_{\dot{W}_{max}} = (0.85 \times 0.90 \times 4)^{\frac{1.4}{2(1.4-1)}} = 7.1$$

The maximum normalized power output at the optimum pressure ratio of 7.1 is calculated using Eq. (7.10).

$$W^* = \frac{\dot{W}_{max}}{\dot{m}c_pT_1} = 0.90 \times 4\left(1 - \frac{1}{\sqrt{0.85 \times 0.90 \times 4}}\right)^2 = 0.661$$

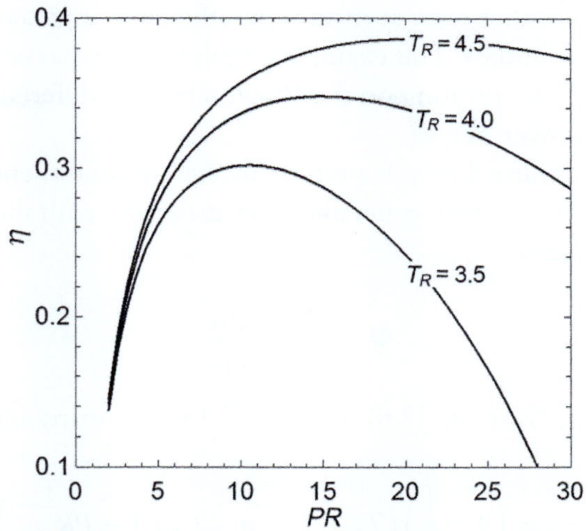

Fig. 7.3 Variation of the thermal efficiency of the irreversible Brayton cycle with the pressure ratio, $\gamma=1.4$, $\eta_{com}=0.85$, and $\eta_{exp}=0.90$.

Fig. 7.3 depicts the thermal efficiency of the Brayton cycle varying with the pressure ratio. Like the power output, the thermal efficiency attains a maximum value at an optimum pressure ratio, which may be obtained by solving $\partial\eta/\partial PR=0$. Hence,

$$(PR)_{\eta_{max}}=\left[\frac{\alpha-\sqrt{\alpha-\beta(\alpha-1)}}{\alpha-1}\right]^{\frac{\gamma}{\gamma-1}} \tag{7.11}$$

where

$$\alpha=\frac{\eta_{exp}T_R}{T_R-1} \tag{7.12}$$

$$\beta=\eta_{com}\eta_{exp}T_R \tag{7.13}$$

As an example, using $\gamma=1.4$, $\eta_{com}=0.85$, $\eta_{exp}=0.90$, and $T_R=4$, we find

$$\alpha=\frac{0.90\times 4}{4-1}=1.2$$
$$\beta=0.85\times 0.90\times 4=3.06$$

$$(PR)_{\eta_{max}}=\left[\frac{1.2-\sqrt{1.2-3.06(1.2-1)}}{1.2-1}\right]^{\frac{1.4}{1.4-1}}=14.95$$

The optimum pressure ratio at maximum efficiency is greater than that at maximum power output. This example reveals that the maximum efficiency design of an engine operating on the Brayton cycle is different from that of the maximum power output.

Now, we examine the cycle performance at minimum entropy generation rate. The total entropy generation rate associated with the operation of the Brayton cycle is

$$\dot{\Phi} = \frac{\dot{Q}_L}{T_L} - \frac{\dot{Q}_H}{T_H} \tag{7.14}$$

Substituting Eqs. (7.5) and (7.6) into Eq. (7.14) and rearranging yields

$$\Phi^* = \frac{\dot{\Phi}}{\dot{m}c_p(T_1/T_L)} = \left(1 - \frac{1}{\pi}\right)(T_R - 1) - \eta_{exp} T_R \left(1 - PR^{\frac{1-\gamma}{\gamma}}\right) + \frac{PR^{\frac{\gamma-1}{\gamma}} - 1}{\pi \eta_{com}} \tag{7.15}$$

where Φ^* denotes the normalized entropy generation and $\pi = T_H/T_L$ is the ratio of the thermal reservoirs' temperatures.

Minimization of the entropy generation given in Eq. (7.15) with respect to the pressure ratio yields

$$(PR)_{\dot{\Phi}_{min}} = \left(\eta_{com}\eta_{exp} T_R \pi\right)^{\frac{\gamma}{2(\gamma-1)}} = (PR)_{\dot{W}_{max}} (\pi)^{\frac{\gamma}{2\gamma-2}} \tag{7.16}$$

Comparing Eq. (7.16) with Eqs. (7.9) and (7.11) reveals that a design based on the minimization of the entropy generation rate is neither equivalent to that of maximum power output nor to that of maximum thermal efficiency. Another subtle observation is that the power output of the engine would be negative if it operated at $(PR)_{\dot{\Phi}_{min}}$. To prove this, recall that the power output of the engine is zero at $PR = \left(\eta_{com}\eta_{exp} T_R\right)^{\frac{\gamma}{\gamma-1}}$.

$$(PR)_{\dot{W}_{net}=0} = \left(\eta_{com}\eta_{exp} T_R\right)^{\frac{\gamma}{\gamma-1}} = \left[(PR)_{\dot{W}_{max}}\right]^2 \tag{7.17}$$

On the other hand, $\pi > T_R$ and $\eta_{com}\eta_{exp} < 1$. So, one may write

$$\pi > \eta_{com}\eta_{exp} T_R \tag{7.18}$$

Using Eq. (7.9), the inequality (7.18) can be rewritten as $\pi > \left[(PR)_{\dot{W}_{max}}\right]^{\frac{2(\gamma-1)}{\gamma}}$, or

$$\pi^{\frac{\gamma}{2(\gamma-1)}} > (PR)_{\dot{W}_{max}} \tag{7.19}$$

Multiplying both sides of the inequality (7.19) by $(PR)_{\dot{W}_{max}}$ gives

$$(PR)_{\dot{W}_{max}} \pi^{\frac{\gamma}{2(\gamma-1)}} > \left[(PR)_{\dot{W}_{max}}\right]^2 \tag{7.20}$$

With the use of Eqs. (7.16) and (7.17), we conclude that

$$(PR)_{\dot{\Phi}_{min}} > (PR)_{\dot{W}_{net}=0} \tag{7.21}$$

Because the optimum pressure ratio corresponding to the minimum entropy generation rate is greater than the pressure ratio at which the power output is zero, the operation of the engine at minimum entropy production would yield a negative power.

7.3 Otto cycle

In the Otto cycle, the heat addition and removal processes take place at constant volume. The rates of heat transferred to and rejected from the air are obtained as

$$\dot{Q}_H = \dot{m}c_v(T_3 - T_2) = \dot{m}c_v T_1\left(T_R - 1 - \frac{CR^{\gamma-1} - 1}{\eta_{com}}\right) \tag{7.22}$$

$$\dot{Q}_L = \dot{m}c_v(T_4 - T_1) = \dot{m}c_v T_1\left[T_R - 1 - \eta_{exp} T_R\left(1 - CR^{1-\gamma}\right)\right] \tag{7.23}$$

where $CR = V_1/V_2$ denotes the compression ratio and the temperature at states 2 and 4 are determined using the following equations.

$$T_2 = T_1\left(1 + \frac{CR^{\gamma-1} - 1}{\eta_{com}}\right) \tag{7.24}$$

$$T_4 = T_3\left[1 - \eta_{exp}\left(1 - CR^{1-\gamma}\right)\right] \tag{7.25}$$

The net power production and the thermal efficiency of the cycle are obtained as follows.

$$\dot{W}_{net} = \dot{Q}_H - \dot{Q}_L = \dot{m}c_v T_1\left(1 - CR^{1-\gamma}\right)\left(\eta_{exp} T_R - \frac{CR^{\gamma-1}}{\eta_{com}}\right) \tag{7.26}$$

$$\eta = \frac{\dot{W}_{net}}{\dot{Q}_H} = \frac{(1 - CR^{1-\gamma})\left(\eta_{exp} T_R - \dfrac{CR^{\gamma-1}}{\eta_{com}}\right)}{T_R - 1 - \dfrac{CR^{\gamma-1} - 1}{\eta_{com}}} \tag{7.27}$$

Fig. 7.4 shows the normalized power and the efficiency of the Otto cycle varying with the compression ratio at $\gamma = 1.4$, $\eta_{com} = 0.85$ and $\eta_{exp} = 0.90$. Like the Brayton cycle, the maximum power design is different from that of the maximum efficiency. The optimum compression ratio leading to a maximum power output happens to be less than that which yields a maximum thermal efficiency. Solving $\partial \dot{W}_{net}/\partial CR = 0$ and $\partial \eta/\partial CR = 0$, one obtains

$$(CR)_{\dot{W}_{max}} = \left(\eta_{com}\eta_{exp} T_R\right)^{\frac{1}{2(\gamma-1)}} \tag{7.28}$$

$$(CR)_{\eta_{max}} = \left[\frac{\alpha - \sqrt{\alpha - \beta(\alpha - 1)}}{\alpha - 1}\right]^{\frac{1}{\gamma-1}} \tag{7.29}$$

where α and β are already defined in Eqs. (7.12) and (7.13).

A substitution of Eq. (7.28) into Eq. (7.26) leads to an expression for the maximum power.

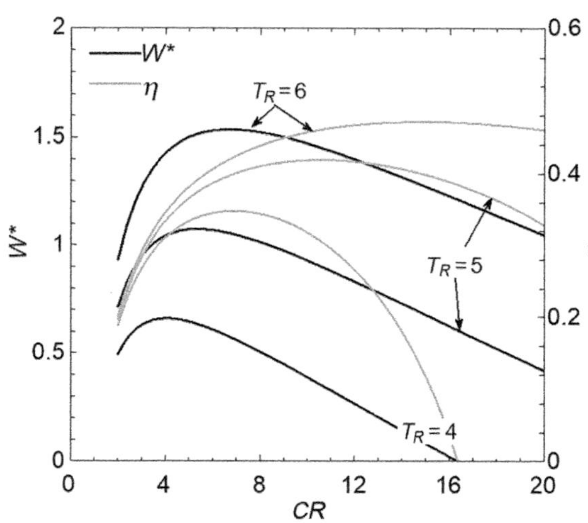

Fig. 7.4 Variation of the normalized power output and the thermal efficiency of the irreversible Otto cycle with the compression ratio, $\gamma = 1.4$, $\eta_{com} = 0.85$, and $\eta_{exp} = 0.90$.

$$\dot{W}_{max} = \dot{m}c_v T_1 \eta_{exp} T_R \left(1 - \frac{1}{\sqrt{\eta_{com}\eta_{exp} T_R}}\right)^2 \tag{7.30}$$

The rate of total entropy generation can be determined by Eq. (7.14) with the use of Eqs. (7.22) and (7.23). Hence,

$$\Phi^* = \frac{\dot{\Phi}}{\dot{m}c_v (T_1/T_L)}$$

$$= \left(1 - \frac{1}{\pi}\right)(T_R - 1) - \eta_{exp} T_R \left(1 - CR^{1-\gamma}\right) + \frac{CR^{\gamma-1} - 1}{\pi \eta_{com}} \tag{7.31}$$

Minimization of Φ^* with respect to the compression ratio gives

$$(CR)_{\dot{\Phi}_{min}} = \left(\eta_{com}\eta_{exp} T_R \pi\right)^{\frac{1}{2(\gamma-1)}} = (CR)_{\dot{W}_{max}} (\pi)^{\frac{1}{2\gamma-2}} \tag{7.32}$$

The optimum pressure given by Eq. (7.32) is different from those of the maximum power output and maximum thermal efficiency, which means that the operational regime at minimum entropy generation rate is not the same as that of the maximum power, nor that of the maximum efficiency. Furthermore, it can be shown with a similar procedure discussed in Section 7.2 for the Brayton cycle that the power output of the Otto cycle would be negative at minimum entropy generation rate. Thus, it is not desirable to operate the engine at minimum entropy generation.

7.4 Atkinson cycle

In the Atkinson cycle, the heat addition process takes place at constant volume through line 2→3 in Fig. 7.1 whereas the heat removal process is isobaric (line 4→1 in Fig. 7.1). Consistent with the analyses of Sections 7.2 and 7.3, expressions should be derived for T_2 and T_4. In Chapter 5, we found relations for the temperature at states 2 and 4 at the condition of isentropic compression and expansion. Here, Eqs. (5.20) and (5.21) are used together with Eqs. (7.1) and (7.2) to determine T_2 and T_4 at nonisentropic condition. Hence,

$$T_2 = T_1 \left[1 + \frac{\left(\frac{T_R}{CR^\gamma}\right)^{1-\gamma} - 1}{\eta_{com}}\right] \tag{7.33}$$

$$T_4 = T_3\left[1 - \eta_{exp}\left(1 - CR^{1-\gamma}\right)\right] \tag{7.34}$$

where $CR = V_4/V_3$ consistent with the definition of the compression ratio; see Section 5.2.4.

The rate of heat transferred from the high-temperature reservoir to the air may now be obtained as follows.

$$\dot{Q}_H = \dot{m}c_v(T_3 - T_2) = \dot{m}c_v T_1\left[T_R - 1 - \frac{\left(\dfrac{T_R}{CR^\gamma}\right)^{1-\gamma} - 1}{\eta_{com}}\right] \tag{7.35}$$

Likewise, an expression can be derived for the rate of heat rejected to the low-temperature reservoir.

$$\dot{Q}_L = \dot{m}c_p(T_4 - T_1) = \dot{m}c_p T_1\left[T_R - 1 - \eta_{exp} T_R\left(1 - CR^{1-\gamma}\right)\right] \tag{7.36}$$

The net power produced by the cycle is the difference between $\dot{Q}_H$ and $\dot{Q}_L$. Hence,

$$\dot{W}_{net} = \dot{m}c_v T_1\left[\gamma\eta_{exp} T_R\left(1 - CR^{1-\gamma}\right) - \frac{\left(\dfrac{T_R}{CR^\gamma}\right)^{1-\gamma} - 1}{\eta_{com}} - (T_R - 1)(\gamma - 1)\right] \tag{7.37}$$

The thermal efficiency of the cycle can be obtained using Eqs. (7.35) and (7.37).

$$\eta = \frac{\gamma\eta_{exp} T_R(1 - CR^{1-\gamma}) - \dfrac{\left(\dfrac{T_R}{CR^\gamma}\right)^{1-\gamma} - 1}{\eta_{com}} - (T_R - 1)(\gamma - 1)}{T_R - 1 - \dfrac{\left(\dfrac{T_R}{CR^\gamma}\right)^{1-\gamma} - 1}{\eta_{com}}} \tag{7.38}$$

Illustrative numerical results are presented in Fig. 7.5 for the normalized power output and the thermal efficiency of the Atkinson cycle varying with the compression ratio. Like the Brayton and Otto cycles, the maximum

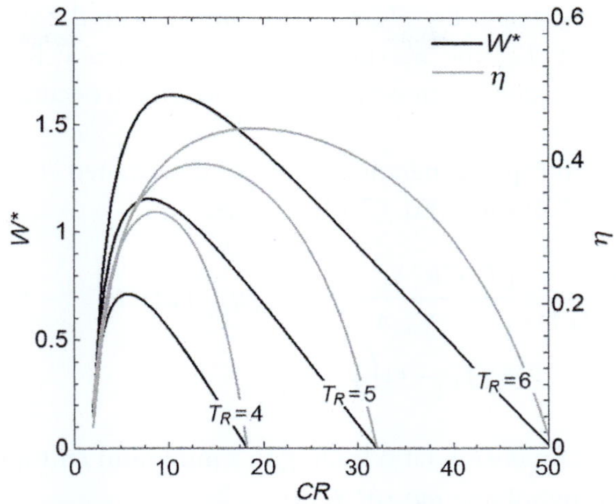

Fig. 7.5 Variation of the normalized power output and the thermal efficiency of the irreversible Atkinson cycle with the compression ratio, $\gamma = 1.4$, $\eta_{com} = 0.85$, and $\eta_{exp} = 0.90$.

power output and the maximum thermal efficiency occur at two different compression ratios. Also, as depicted in Fig. 7.5, the compression ratio leading to the maximum power is less than that of the maximum efficiency under identical design condition. Solving $\partial \dot{W}_{net}/\partial CR = 0$, we find

$$(CR)_{\dot{W}_{max}} = \left(\eta_{com}\eta_{exp}T_R^{\gamma}\right)^{\frac{1}{\gamma^2-1}} \tag{7.39}$$

For instance, substitute $\gamma = 1.4$, $\eta_{com} = 0.85$, and $\eta_{exp} = 0.90$ into Eq. (7.39). The optimum compression ratio at $T_R = 4$ is found as

$$(CR)_{\dot{W}_{max}} = \left(0.85 \times 0.90 \times 4^{1.4}\right)^{\frac{1}{1.4^2-1}} = 5.7$$

The normalized maximum power output at the compression ratio of 5.7 is computed using Eq. (7.37). Hence,

$$\frac{\dot{W}_{max}}{\dot{m}c_v T_1} = 1.4 \times 0.9 \times 4\left(1 - 5.7^{-0.4}\right) - \frac{\left(\frac{4}{5.7^{1.4}}\right)^{-0.4} - 1}{0.85} - (4-1)(1.4-1)$$
$$= 0.713$$

The results can be verified in Fig. 7.5. For the optimum compression ratio leading to a maximum thermal efficiency, there is no suitable analytical expression. So, one would need to numerically search for the extremum of the efficiency given in Eq. (7.38). For example, with the above values of γ, η_{com}, η_{exp}, and T_R, the maximum thermal efficiency is found to be 0.328 using the Golden-Section Search method, which occurs at a compression ratio of 8.6.

The total entropy generation rate is determined by substituting Eqs. (7.35) and (7.36) into Eq. (7.14). Hence,

$$\frac{\dot{\Phi}}{\dot{m}c_v(T_1/T_L)} = \frac{\left(T_R CR^{-\gamma}\right)^{1-\gamma} - 1}{\eta_{com}\pi} + \gamma\eta_{exp}T_R\left(CR^{1-\gamma} - 1\right) + (T_R - 1)\left(\gamma - \frac{1}{\pi}\right) \tag{7.40}$$

The optimum compression ratio leading to a minimum entropy generation rate is obtained by solving $\partial\dot{\Phi}/\partial CR = 0$.

$$(CR)_{\dot{\Phi}_{min}} = \left(\eta_{com}\eta_{exp}T_R^{\gamma}\pi\right)^{\frac{1}{\gamma^2-1}} = (CR)_{\dot{W}_{max}}(\pi)^{\frac{1}{\gamma^2-1}} \tag{7.41}$$

For $\gamma = 1.4$, we have $\frac{1}{\gamma^2-1} \sim 1$. So, it can be deduced from Eq. (7.41) that the compression ratio at minimum entropy generation rate is about π times greater than that at maximum power output. Fig. 7.6 shows an illustrative example assuming $\pi = 5$ and using the same values of γ, η_{com}, η_{exp}, and T_R as before. The optimum compression ratio maximizing the power output was determined to be 5.7. Using Eq. (7.41), we have

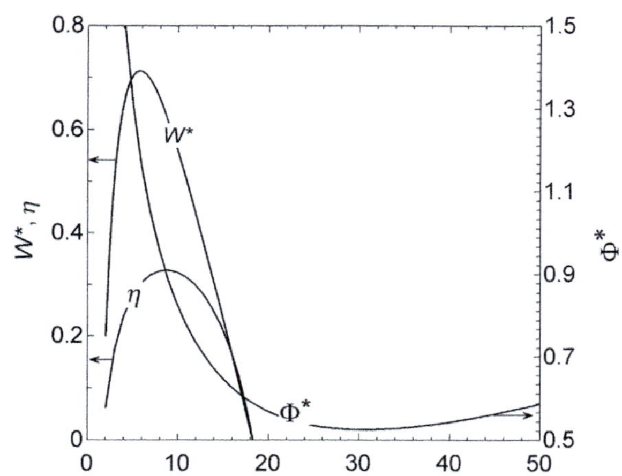

Fig. 7.6 The normalized power output, thermal efficiency, and normalized entropy generation rate of the irreversible Atkinson cycle varying with the compression ratio, $T_R = 4$ and $\pi = 5$.

$$(CR)_{\dot{\Phi}_{min}} = (CR)_{\dot{W}_{max}} (\pi)^{\frac{1}{\gamma^2 - 1}} \sim (CR)_{\dot{W}_{max}} \pi = (5.7)(5) = 28.5$$

The optimum compression ratio minimizing the entropy generation rate is 28.5, which could also be verified in Fig. 7.6. However, this compression ratio is impractical since both the power and the efficiency of the engine approach zero at a compression ratio of 18.

7.5 Diesel cycle

In Diesel cycle, the heat addition and heat removal processes take place at constant pressure and constant volume through lines 2→3 and 4→1 in Fig. 7.1. The air temperature at states 2 and 4 can be obtained using the following relations.

$$T_2 = T_1 \left(1 + \frac{CR^{\gamma-1} - 1}{\eta_{com}} \right) \tag{7.42}$$

$$T_4 = T_3 \left\{ 1 - \eta_{exp} \left[1 - \left(\frac{CR^{\gamma}}{T_R} \right)^{1-\gamma} \right] \right\} \tag{7.43}$$

where $CR = V_1/V_2$.

The heat transfer rates between the thermal reservoirs and the engine are obtained as follows.

$$\dot{Q}_H = \dot{m}c_p(T_3 - T_2) = \dot{m}c_p T_1 \left(T_R - 1 - \frac{CR^{\gamma-1} - 1}{\eta_{com}} \right) \tag{7.44}$$

$$\dot{Q}_L = \dot{m}c_v(T_4 - T_1) = \dot{m}c_v T_1 \left\{ T_R - 1 - \eta_{exp} T_R \left[1 - \left(\frac{CR^{\gamma}}{T_R} \right)^{1-\gamma} \right] \right\} \tag{7.45}$$

Thus, the expressions for the net power production and the thermal efficiency read

$$\dot{W}_{net} = \dot{m}c_v T_1 \left\{ \eta_{exp} T_R \left[1 - \left(\frac{CR^{\gamma}}{T_R} \right)^{1-\gamma} \right] - \gamma \frac{CR^{\gamma-1} - 1}{\eta_{com}} + (T_R - 1)(\gamma - 1) \right\} \tag{7.46}$$

$$\eta = \frac{\eta_{exp} T_R \left[1 - \left(\frac{CR^{\gamma}}{T_R}\right)^{1-\gamma}\right] - \gamma \frac{CR^{\gamma-1} - 1}{\eta_{com}} + (T_R - 1)(\gamma - 1)}{\gamma\left(T_R - 1 - \frac{CR^{\gamma-1} - 1}{\eta_{com}}\right)} \tag{7.47}$$

Maximizing the net power of the engine with respect to the compression ratio, one obtains

$$(CR)_{\dot{W}_{max}} = \left(\eta_{com}\eta_{exp} T_R^{\gamma}\right)^{\frac{1}{\gamma^2-1}}$$

which is exactly the same result that we found for the Atkinson cycle; see Eq. (7.39). In other words, for identical values of γ, η_{com}, η_{exp}, and T_R, both Atkinson and Diesel cycles attain a maximum power at the same compression ratio. For example, in the preceding section, we found an optimum compression ratio of 5.7 for the Atkinson cycle at maximum power for $\gamma = 1.4$, $\eta_{com} = 0.85$, $\eta_{exp} = 0.90$, and $T_R = 4$. Substituting $CR = 5.7$ into Eq. (7.46) gives the maximum normalized power for the Diesel cycle.

$$\begin{aligned} \frac{\dot{W}_{max}}{\dot{m} c_v T_1} &= 0.9 \times 4\left[1 - \left(\frac{5.7^{1.4}}{4}\right)^{-0.4}\right] - (1.4)\frac{5.7^{0.4} - 1}{0.85} + (4-1)(1.4-1) \\ &= 0.778 \end{aligned}$$

Like the Atkinson cycle, the optimum compression ratio maximizing the thermal efficiency of the Diesel cycle needs to be found through numerical methods. Using the above values of γ, η_{com}, η_{exp}, and T_R, a Golden-Section Search method gives a maximum efficiency of 0.345 that occurs at an optimum compression ratio of 8.9.

The following expression is also resulted for the total entropy generation rate of the cycle.

$$\begin{aligned} \frac{\dot{\Phi}}{\dot{m} c_v (T_1/T_L)} = &\gamma \frac{CR^{\gamma-1} - 1}{\eta_{com}\pi} - \eta_{exp} T_R \left[1 - \left(\frac{CR^{\gamma}}{T_R}\right)^{1-\gamma}\right] \\ &+ (T_R - 1)\left(1 - \frac{\gamma}{\pi}\right) \end{aligned} \tag{7.48}$$

Applying $\partial\dot{\Phi}/\partial CR = 0$, we get an expression for the optimum compression ratio that minimizes the entropy generation rate.

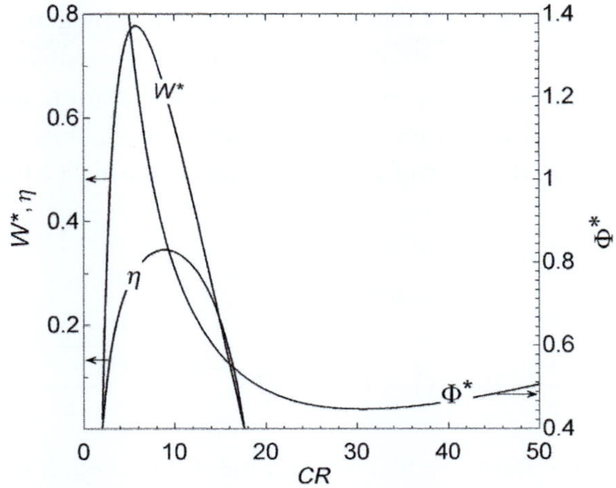

Fig. 7.7 The normalized power output, thermal efficiency, and normalized entropy generation rate of the irreversible Diesel cycle varying with the compression ratio, $\gamma = 1.4$, $\eta_{com} = 0.85$, $\eta_{exp} = 0.90$, $T_R = 4$, and $\pi = 5$.

$$(CR)_{\dot{\Phi}_{min}} = (CR)_{\dot{W}_{max}} (\pi)^{\frac{1}{\gamma^2 - 1}}$$

which is the same result that we found for the Atkinson cycle; see Eq. (7.41). Fig. 7.7 depicts an illustrative example where the normalized power output, thermal efficiency, and normalized entropy generation rate of the Diesel cycle are plotted against the compression ratio. Evident from Fig. 7.7 is that a design based on minimization of entropy generation rate would be impractical, consistent with the observations we made previously for the Brayton, Otto, and Atkinson cycles.

7.6 Isentropic compression and expansion

A special case worthy of discussion from a theoretical perspective is that the isentropic efficiencies of the compression and expansion processes are assumed to be 100%, i.e., $\eta_{com} = \eta_{exp} = 1$. In this case, the irreversible models of the Brayton, Otto, Atkinson, and Diesel cycles reduce to their endoreversible designs presented in Chapter 5. The only irreversibility would then be due to the heat exchange between the thermal reservoirs and the engine. In this case, the thermal efficiency would then be determined using Eqs. (5.9), (5.14), (5.22), and (5.27) for the Brayton, Otto, Atkinson, Diesel cycles, respectively.

In Chapter 6, we learned that a reduction in entropy generation corresponds to an increase in thermal efficiency in endoreversible engines. We now examine the efficiency-entropy production relation in the power cycles discussed in this chapter at the condition of $\eta_{com}=\eta_{exp}=1$. For the ideal Brayton cycle, the thermal efficiency and the normalized entropy generation read

$$\eta = 1 - PR^{\frac{1-\gamma}{\gamma}} \tag{7.49}$$

$$\Phi^* = \left(1-\frac{1}{\pi}\right)(T_R-1) - T_R\left(1-PR^{\frac{1-\gamma}{\gamma}}\right) + \frac{PR^{\frac{\gamma-1}{\gamma}}-1}{\pi} \tag{7.50}$$

Fig. 7.8 shows the thermal efficiency and the normalized entropy production of the ideal Brayton cycle at $T_R=4$ and $\pi=5$. The efficiency monotonically increases with the pressure ratio, whereas the entropy generation decreases with an increase in the pressure ratio. It can be shown mathematically that the inverse relation of the thermal efficiency with the entropy production is always valid for the ideal Brayton cycle. A combination of Eqs. (7.49) and (7.50) to eliminate $PR^{(\gamma-1)/\gamma}$ yields

$$\Phi^* = \left(1-\frac{1}{\pi}\right)(T_R-1) - T_R\eta + \frac{1}{\pi}\frac{\eta}{1-\eta} \tag{7.51}$$

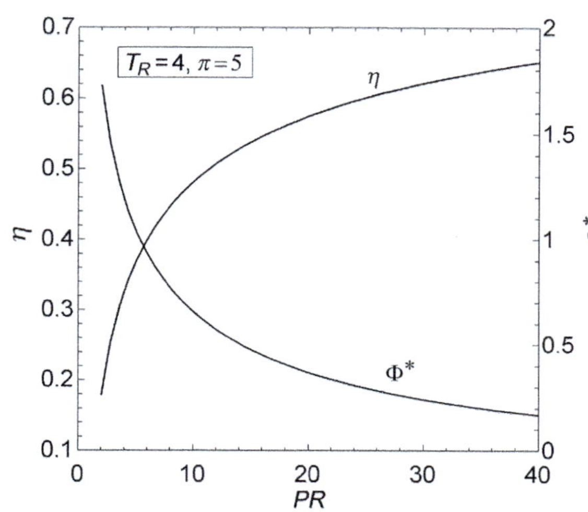

Fig. 7.8 An illustrative example showing the inverse relation between the thermal efficiency and the entropy generation in an *ideal* Brayton cycle.

Differentiating Φ^* with respect to η, we get

$$\frac{d\Phi^*}{d\eta} = -T_R + \frac{1}{\pi(1-\eta)^2} \tag{7.52}$$

We only need to show that the right-hand side of Eq. (7.52) is always negative. For this, we use Eq. (5.7), where $1-\eta = T_1/T_2$. Also, because $T_R < \pi$, we may write

$$-T_R + \frac{1}{\pi(1-\eta)^2} < -T_R + \frac{1}{T_R(1-\eta)^2} = -\frac{1}{T_R}\left[T_R^2 - \left(\frac{T_2}{T_1}\right)^2\right]$$
$$= -\frac{1}{T_R}\left(\frac{T_3^2 - T_2^2}{T_1^2}\right) < 0$$

We therefore conclude that in an ideal Brayton cycle $d\Phi^*/d\eta < 0$. In other words, an increase in the thermal efficiency corresponds to a decrease in the entropy generation rate in an *ideal* Brayton cycle. Similar analysis can be made to show that in ideal Otto, Atkinson, and Diesel cycles, the thermal efficiency correlates inversely with the entropy generation rate. This is graphically demonstrated in Fig. 7.9.

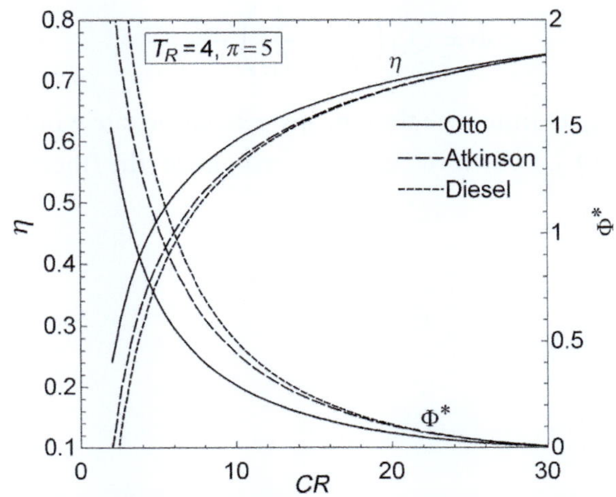

Fig. 7.9 Graphical presentation of the inverse relation between the thermal efficiency and the entropy generation in *ideal* Otto, Atkinson, and Diesel cycles.

7.7 Fixed heat input

Another special case is that the quantity of heat supplied to engine is fixed—a design constraint. For an engine operating in a closed cycle, which receives heat from a single high-temperature reservoir, and it rejects the unused amount of heat to a low-temperature reservoir, it can be readily shown that the minimization of entropy generation rate, maximization of power output, and maximization of thermal efficiency are all equivalent.

The net power production is the difference between the rates of heat supplied to/rejected from the engine, i.e., $\dot{W}_{net} = \dot{Q}_H - \dot{Q}_L$. For a given $\dot{Q}_H$, the net power is maximized by minimizing the rate of heat rejected from the cycle $\dot{Q}_L$. From the definition of the thermal efficiency, $\eta = \dot{W}_{net}/\dot{Q}_H$, maximization of efficiency would be equivalent to maximization of the power when $\dot{Q}_H$ is treated as a constant parameter. On the other hand, the rate of total entropy generation given by Eq. (7.14) is minimized by minimization of the $\dot{Q}_L$. Thus, the three optimization criteria would be identical at the condition of fixed heat input.

One may argue with a similar reasoning that in an engine that exchanges heat with two thermal reservoirs, maximization of thermal efficiency and minimization of the total entropy generation rate would yield an identical design at the condition of fixed power output $\dot{W}_{net}$. Eliminating $\dot{Q}_L$ between $\dot{W}_{net} = \dot{Q}_H - \dot{Q}_L$ and Eq. (7.14), we get

$$\dot{\Phi} = \dot{Q}_H\left(\frac{1}{T_L} - \frac{1}{T_H}\right) - \frac{\dot{W}_{net}}{T_L} \tag{7.53}$$

For a fixed $\dot{W}_{net}$, to minimize the entropy generation rate would require one to minimize $\dot{Q}_H$, which would also maximize the thermal efficiency $\eta = \dot{W}_{net}/\dot{Q}_H$.

CHAPTER EIGHT

Irreversible engines—Open cycles

8.1 Introduction

The conventional methods for production of power require combustion of fuel in air. The energy content of the fuel is then released in the form of heat, which is transferred to the working substance either directly (e.g., internal combustion engines) or indirectly (e.g., steam power cycle). Combustion-driven power generation systems (e.g., gas turbines cycles, steam or combined cycle power plants, and internal combustion engines) have one feature in common in that fuel and air are supplied through two different streams, which are then mixed and reacted. A portion of the combustion heat is wasted and discharged to the surrounding. For example, in a steam power cycle (Fig. 8.1), the waste heat is rejected to the environment through the condenser as well as the flue gases that leave the stack of the boiler at a moderate temperature (e.g., 413 K).

As discussed in Chapter 7, minimizing entropy generation rate may only be equivalent to maximizing thermal efficiency under certain design conditions such as fixed rate of heat input or fixed power output. To rectify this issue, we introduce a new parameter referred to as specific entropy generation (SEG) that is defined as the rate of total entropy generation per unit flowrate of the fuel combusted, and has units of J/(mol K). It will be shown that optimizing a power system for maximum thermal efficiency is always equivalent to optimizing for minimum SEG. The SEG method will then be applied to typical combustion-driven power cycles, including gas turbine and combined cycles.

8.2 Specific entropy generation

Consider a power-generating system, which takes in a hydrocarbon fuel. Assume that both fuel and air are supplied at the ambient temperature and pressure (T_0, p_0). Denoting the net power production by $\dot{W}_{net}$ and the

Entropy Analysis in Thermal Engineering Systems
https://doi.org/10.1016/B978-0-12-819168-2.00008-8

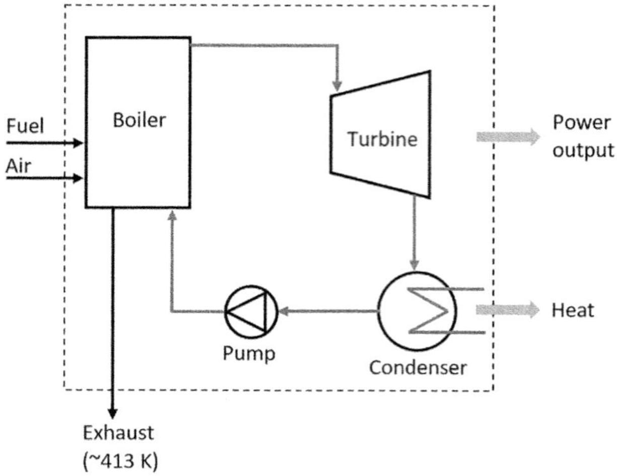

Fig. 8.1 Schematic of a steam power cycle. The dashed rectangle shows the system boundary.

rate of waste heat discharged to the atmosphere by $\dot{Q}_0$, the conservation of energy equation applied to the system is

$$H_0^a + H_0^f = \dot{W}_{net} + \dot{Q}_0 + H_0^p \tag{8.1}$$

where H_0^a is the enthalpy of air, H_0^f the enthalpy of fuel, and H_0^p the enthalpy of the combustion products at ambient conditions. Note that H accounts for both the enthalpy of formation and the sensible enthalpy.

The total entropy generation rate of the power-generating system is determined by accounting for the entropy generation within the system, $\langle\dot{\Phi}\rangle_{system}$, and that due to the waste heat rejected to the surroundings, $\langle\dot{\Phi}\rangle_{wasteheat}$. Hence, the total rate of entropy generation is

$$\dot{\Phi} = \langle\dot{\Phi}\rangle_{system} + \langle\dot{\Phi}\rangle_{wasteheat} = \left(S_e^p - S_0^a - S_0^f\right) + \left(\frac{\dot{Q}_0}{T_0} + S_{p,0} - S_e^p\right) \tag{8.2}$$

where S_0^a and S_0^f denote the entropy flows of the ambient air and fuel into the system, S_e^p is the entropy flow of the reaction products at the exit of the system, and S_0^p denotes the entropy flow of the products at the ambient temperature and pressure. When evaluating flow entropies, the partial pressure of each substance in the reactants and products streams must be accurately assessed. For instance, the entropy of carbon dioxide as a by-product of fuel oxidation should be calculated based on its concentration in the combustion products stream.

A combination of Eqs. (8.1) and (8.2) to eliminate $\dot{Q}_0$ yields

$$\dot{W}_{net} = \left(H_0^a + H_0^f - H_0^p\right) + T_0\left(S_0^p - S_0^a - S_0^f\right) - T_0\dot{\Phi} \qquad (8.3)$$

which reduces to the following expression if the system is fully reversible

$$\dot{W}_{rev} = \left(H_0^a + H_0^f - H_0^p\right) + T_0\left(S_0^p - S_0^a - S_0^f\right) \qquad (8.4)$$

If Eq. (8.3) is described per unit molar flowrate of fuel burned $\dot{n}_f$, we have

$$w_{net} = w_{rev} - T_0\mathrm{SEG} \qquad (8.5)$$

where $\mathrm{SEG} = \dot{\Phi}/\dot{n}_f$ denotes the specific entropy generation, and

$$w_{rev} = \frac{1}{\dot{n}_f}\left(H_0^a + H_0^f - H_0^p\right) + \frac{T_0}{\dot{n}_f}\left(S_0^p - S_0^a - S_0^f\right) \qquad (8.6)$$

In the next section, we will mathematically show that the specific reversible work, w_{rev}, is approximately a fixed quantity under different operational conditions. In order to reach this goal, the enthalpy and entropy terms in Eq. (8.6) must be evaluated for a generic combustion reaction.

8.3 Proof that w_{rev} is relatively constant

The proof to be presented is based on the following set of common assumptions: (i) air consists of 21% oxygen and 79% nitrogen (molar basis), (ii) fuel contains only carbon, hydrogen, and oxygen, to be represented by $C_xH_yO_z$, (iii) air and the combustion products behave as ideal gas mixtures, and (iv) the combustion products consist of carbon dioxide, water, oxygen, and nitrogen. As Eq. (8.6) requires calculation of the enthalpy and entropy of the products at ambient conditions, water vapor condensation is also accounted for in the products stream. The overall combustion reaction can be represented as

$$C_xH_yO_z + \Lambda(O_2 + 3.76N_2) \rightarrow xCO_2 + n_vH_2O_{(v)} + \left(\Lambda - x - \frac{y}{4} + \frac{z}{2}\right)O_2$$
$$+3.76\Lambda N_2 + \left(\frac{y}{2} - n_v\right)H_2O_{(l)}$$

where Λ is a stochiometric coefficient. The number of moles of air supplied to combust one mole of fuel is 4.76Λ. The minimum value of Λ corresponds to stoichiometric combustion: $\Lambda_{\min} = x + \frac{y}{4} - \frac{z}{2}$. Also, n_v denotes the number of moles of water vapor in the gaseous products, which depends on its partial pressure [1].

$$\frac{n_v}{n_p} = \frac{p_{sat}(T_0)}{p_0} \tag{8.7}$$

where $p_{sat}(T_0)$ is the saturation pressure of water at ambient temperature, and n_p is the total number of moles of combustion products in the gas phase.

$$n_p = n_v - \frac{y}{4} + \frac{z}{2} + 4.76\Lambda \tag{8.8}$$

The combustion of one mole of fuel with 4.76Λ moles of air yields n_p moles of gaseous products and $\left(\frac{y}{2} - n_v\right)$ moles of liquid water. The molar enthalpies and entropies of the air, fuel, and combustion products are determined per unit mole of the fuel.

$$\begin{aligned}\frac{1}{\dot{n}_f}\left(H_0^a + H_0^f - H_0^p\right) &= \left(\Lambda h_{O_2} + 3.76\Lambda h_{N_2}\right)_0^a + h_{f,0} - \left(\frac{y}{2} - n_v\right) h_{H_2O_{(l)},0} \\ &- \left[x h_{CO_2} + n_v h_{H_2O_{(v)}} + \left(\Lambda - x - \frac{y}{4} + \frac{z}{2}\right) h_{O_2} + 3.76\Lambda h_{N_2}\right]_0^p\end{aligned} \tag{8.9}$$

Recalling that both the air and fuel are supplied to the system at ambient conditions, Eq. (8.9) can be simplified to

$$\begin{aligned}\frac{1}{\dot{n}_f}\left(H_0^a + H_0^f - H_0^p\right) &= \left[h_{f,0} - x h_{CO_2,0} - \left(\frac{y}{2}\right) h_{H_2O_{(l)},0}\right] + \left(x + \frac{y}{4} - \frac{z}{2}\right) h_{O_2,0} \\ &- n_v \left[h_{H_2O_{(v)},0} - h_{H_2O_{(l)},0}\right]\end{aligned} \tag{8.10}$$

Taking the reference temperature and pressure $T_0 = 298.15$ K and $p_0 = 1$ bar, the first three terms on the right-hand-side of Eq. (8.10) constitute the higher heating value (HHV) of the fuel, and the fourth term is zero because $h_{O_2,0} = 0$. The enthalpy difference in the last term is the enthalpy of evaporation of water, $\mathcal{L}_0^w$. Hence, Eq. (8.10) reduces to

$$\frac{1}{\dot{n}_f}\left(H_0^a + H_0^f - H_0^p\right) = \text{HHV} - n_v \mathcal{L}_0^w \tag{8.11}$$

Eq. (8.11) expresses the first right-hand-side term from Eq. (8.6) as a simple function of HHV and the enthalpy of water evaporation. Next, the second right-hand-side term from Eq. (8.6) is determined as follows.

$$\frac{1}{\dot{n}_f}\left(S_0^p - S_0^a - S_0^f\right) = \left[x s_{CO_2} + n_v s_{H_2O_{(v)}} + \left(\Lambda - x - \frac{y}{4} + \frac{z}{2}\right) s_{O_2} + 3.76\Lambda s_{N_2}\right]_0^p - \left(\Lambda s_{O_2} + 3.76\Lambda s_{N_2}\right)_0^a - s_{f,0} + \left(\frac{y}{2} - n_v\right) s_{H_2O_{(l)},0} \tag{8.12}$$

Rearranging Eq. (8.12) to group similar species together yields

$$\frac{1}{\dot{n}_f}\left(S_0^p - S_0^a - S_0^f\right) = \Lambda\left[\left(s_{O_2}\right)_0^p - \left(s_{O_2}\right)_0^a\right] + 3.76\Lambda\left[\left(s_{N_2}\right)_0^p - \left(s_{N_2}\right)_0^a\right] - s_{f,0} + \left(\frac{y}{2} - n_v\right) s_{H_2O_{(l)},0} + \left[x s_{CO_2} + n_v s_{H_2O_{(v)}} + \left(-x - \frac{y}{4} + \frac{z}{2}\right) s_{O_2}\right]_0^p \tag{8.13}$$

where

$$\left(s_{O_2}\right)_0^p - \left(s_{O_2}\right)_0^a = R \ln\left(\frac{y_{O_2}^a}{y_{O_2}^p}\right) \tag{8.14}$$

$$\left(s_{N_2}\right)_0^p - \left(s_{N_2}\right)_0^a = R \ln\left(\frac{y_{N_2}^a}{y_{N_2}^p}\right) \tag{8.15}$$

$$\left(s_{CO_2}\right)_0^p = s_{CO_2,0} - R \ln\left(y_{CO_2}^p\right) \tag{8.16}$$

$$\left(s_{H_2O_{(v)}}\right)_0^p = s_{H_2O_{(v)},0} - R \ln\left(y_{H_2O_{(v)}}^p\right) = s_{H_2O_{(v)},0} - R \ln\left[\frac{p_{sat}(T_0)}{p_0}\right] \tag{8.17}$$

$$\left(s_{O_2}\right)_0^p = s_{O_2,0} - R \ln\left(y_{O_2}^p\right) \tag{8.18}$$

where y_i^j denotes the mole fraction of species i (O_2, N_2, CO_2, H_2O) in mixture j (air, combustion products), R is the universal gas constant, and $s_{i,0}$ is the standard entropy of species i.

Substituting Eqs. (8.14)–(8.18) into Eq. (8.13) and rearranging yields

$$\frac{1}{\dot{n}_f}\left(S_0^p - S_0^a - S_0^f\right) = R\ \ln\left[\frac{\left(y_{O_2}^a\right)^{\Lambda}\left(y_{N_2}^a\right)^{3.76\Lambda}}{\left(y_{N_2}^p\right)^{3.76\Lambda}\left(y_{CO_2}^p\right)^{x}\left(y_{O_2}^p\right)^{\left(\Lambda - x - \frac{y}{4} + \frac{z}{2}\right)}}\right] + \mathcal{C}_1 + n_v\mathcal{C}_2 \tag{8.19}$$

where

$$\mathcal{C}_1 = -s_{f,0} + \left(\frac{y}{2}\right)s_{H_2O_{(l)},0} + x s_{CO_2,0} + \left(-x - \frac{y}{4} + \frac{z}{2}\right)s_{O_2,0} \tag{8.20}$$

$$\mathcal{C}_2 = s_{H_2O_{(v)},0} - s_{H_2O_{(l)},0} - R\ \ln\left[\frac{p_{sat}(T_0)}{p_0}\right] \tag{8.21}$$

Eqs. (8.11) and (8.19) can now be substituted into Eq. (8.6) to determine w_{rev}. Hence,

$$w_{rev} = HHV + T_0\mathcal{C}_1 + n_v\left(T_0\mathcal{C}_2 - \mathcal{L}_0^w\right) + T_0 R\ \ln\left[\frac{\left(y_{O_2}^a\right)^{\Lambda}\left(y_{N_2}^a\right)^{3.76\Lambda}}{\left(y_{N_2}^p\right)^{3.76\Lambda}\left(y_{CO_2}^p\right)^{x}\left(y_{O_2}^p\right)^{\left(\Lambda - x - \frac{y}{4} + \frac{z}{2}\right)}}\right] \tag{8.22}$$

Note that $\mathcal{C}_2$ given in Eq. (8.21) is the entropy change due to the evaporation of water at T_0. Hence

$$\mathcal{C}_2 = \frac{\mathcal{L}_0^w}{T_0} \tag{8.23}$$

At 298.15 K and 1 bar, the entropy of water in gas phase and liquid phase is 188.8J/(mol K) and 69.95J/(mol K), respectively, and its evaporation enthalpy is 44,000J/mol [2]. The saturation pressure at 298.15 K is 0.0317 bar. Substituting these values into Eq. (8.21), we find

$$\mathcal{C}_2 = 188.8 - 69.95 - 8.318\ \ln\left(\frac{0.0317}{1}\right) = 147.55\ (\text{J/mol K})$$

Alternatively, $\mathcal{C}_2$ computed using Eq. (8.23) is

$$\mathcal{C}_2 = \frac{\mathcal{L}_0^w}{T_0} = \frac{44,000}{298.15} = 147.58\ (\text{J/mol K})$$

The two values obtained using Eqs. (8.21) and (8.23) are very close. Thus, from Eq. (8.23), one may conclude that $T_0\mathcal{C}_2 - \mathcal{L}_0^w = 0$. Eq. (8.22) may finally be expressed as

$$w_{rev} = \mathcal{C} + \mathcal{F}(\Lambda) \tag{8.24}$$

where

$$\mathcal{C} = \text{HHV} + T_0\mathcal{C}_1 \tag{8.25}$$

$$\mathcal{F}(\Lambda) = T_0 R \ \ln\left[\frac{\left(y_{O_2}^a\right)^{\Lambda}\left(y_{N_2}^a\right)^{3.76\Lambda}}{\left(y_{N_2}^p\right)^{3.76\Lambda}\left(y_{CO_2}^p\right)^{x}\left(y_{O_2}^p\right)^{\left(\Lambda - x - \frac{y}{4} + \frac{z}{2}\right)}}\right] \tag{8.26}$$

It can be deduced from Eqs. (8.20) and (8.25) that $\mathcal{C}_1$, and therefore $\mathcal{C}$ are constant parameters for a specified fuel at the ambient temperature and pressure. So, w_{rev} can be represented as a function of fuel-air equivalence ratio, ϕ: $w_{rev} = w_{rev}(\phi)$, where $\phi = \Lambda/\Lambda_{min}$.

The numerical values of w_{rev} are calculated and given in Table 8.1 for 13 different fuels over a range of equivalence ratios. The last column, Δw_{rev}, lists the percentage increase in w_{rev} as ϕ is raised from 1 to 5. These results indicate that a fivefold increase in ϕ only raises w_{rev} by 1.7% on average. Thus, w_{rev} appears to be a very weak function of ϕ. Note that for hydrogen and carbon monoxide, w_{rev} is less than the lower heating value (LHV), whereas for other fuels it is between the lower and higher heating values.

The trend observed in Table 8.1 can be written as

$$w_{rev}(\phi_1) \approx w_{rev}(\phi_2) \ldots \approx w_{rev}(\phi_n) \quad 1 \le \phi_i \le 5 \quad i = 1, 2, \ldots, n \tag{8.27}$$

It can be implied from Eq. (8.27) that the numerical value of $\mathcal{F}(\Lambda)$ is negligible compared to constant $\mathcal{C}$ in Eq. (8.24) whereby $w_{rev} \cong \mathcal{C}$. Alternatively, w_{rev} may be expressed in terms of any ϕ within the range $1 \le \phi_i \le 5$. For the sake of simplicity, $\phi_{min} = 1$ is selected. Noting the definition of thermal efficiency $\eta = \frac{w_{net}}{HV}$, Eq. (8.5) results in the following inverse relationship between η and SEG.

$$\eta = \frac{w_{rev}(\Lambda_{min}) - T_o\text{SEG}}{HV} \tag{8.28}$$

where $w_{rev}(\Lambda_{min})$ is evaluated using Eqs. (8.24)–(8.26). Note that both η and SEG are dependent on several process parameters such as the highest temperature/pressure of the working fluid. Because $w_{rev}(\Lambda_{min})$ is constant for a

Table 8.1 Numerical values of w_{rev} for different fuels and fuel-air equivalence ratios.

Fuel	Chemical formula	LHV (kJ/mol)	HHV (kJ/mol)	Λ_{min}	w_{rev} (kJ/mol) $\phi = 1$[a]	$\phi = 3$	$\phi = 5$	Δw_{rev} (%)
Hydrogen	H_2	241.8	285.8	0.5	234.4	235.9	236.7	0.98
Carbon monoxide	CO	283.0[b]	283.0	0.5	259.2	263.5	265.5	2.43
Methanol	CH_3OH	676.2	764.2	1.5	705.2	712.8	716.4	1.59
Methane	CH_4	802.5	890.6	2	814.8	824.0	828.3	1.66
Ethanol	C_2H_5OH	1278	1410	3	1330	1345	1352	1.65
Acetylene	C_2H_2	1257	1301	2.5	1236	1250	1256	1.62
Ethylene	C_2H_4	1323	1411	3	1329	1344	1351	1.66
Ethane	C_2H_6	1429	1561	3.5	1464	1481	1488	1.64
Propane	C_3H_8	2043	2219	5	2102	2126	2138	1.71
Butane	C_4H_{10}	2657	2877	6.5	2740	2772	2787	1.72
Pentane	C_5H_{12}	3272	3536	8	3378	3418	3436	1.72
Hexane	C_6H_{14}	3887	4195	9.5	4017	4064	4086	1.72
Octane	C_8H_{18}	5116	5512	12.5	5294	5356	5386	1.74

All calculations are performed at 298.15 K and 1 bar.

[a] $\phi = \frac{\Lambda}{\Lambda_{min}}$.

[b] CO has only one heating value.

given fuel, Eq. (8.28) proves that the maximum thermal efficiency always corresponds to the minimum SEG with respect to any design or process parameter. This result holds for any combustion power system within the proper range of the equivalence ratio, ϕ.

In the next sections, the inverse relation of the thermal efficiency and specific entropy generation will be demonstrated numerically for typical power plants.

8.4 Gas turbine cycle

Fig. 8.2 shows a schematic of a gas turbine power cycle that operates on irreversible open Brayton cycle. It consists of an air compressor, a fuel compressor, a combustor, and a turbine. In addition to the assumptions made in the beginning of Section 8.3, it will be assumed that all cycle components operate adiabatically (no heat leakage), the pressure drop on the path of the working fluid is negligible, and air enters the compressor at the ambient temperature ($T_1 = T_0$) and pressure ($p_1 = p_0$).

The thermodynamic model of the gas turbine cycle depicted in Fig. 8.2 is similar to that of the irreversible closed Brayton cycle described in

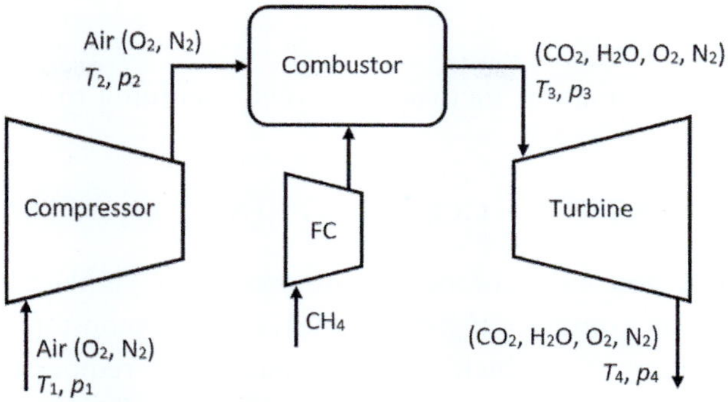

Fig. 8.2 Schematic of a gas turbine cycle.

Chapter 7. Here, we will also account for the temperature dependence of the specific heat of the gaseous species as well as the fuel combustion in the pressurized air coming from the compressor. The fuel is methane (CH_4) that is compressed from the ambient pressure and temperature to the combustor pressure. The analysis begins by calculating the molar enthalpy and entropy flows per unit molar flowrate of the fuel at the inlet and outlet of the components.

8.4.1 Enthalpy and entropy flows

The compressor outlet temperature under isentropic condition is obtained by

$$T_{2s} = T_1 \left(\frac{p_2}{p_1} \right)^{\frac{\gamma_c - 1}{\gamma_c}} \tag{8.29}$$

where γ_c denotes the average specific heat ratio of the air across the compressor.

The isentropic efficiency of the compressor defined in terms of enthalpies as

$$\eta_c = \frac{h_{2s} - h_1}{h_2 - h_1} \tag{8.30}$$

allows us to determine the air temperature at the compressor outlet. The enthalpy and entropy of the air at the entrance and the exit of the compressor are calculated as follows.

$$H_j = \left(\Lambda h_{O_2} + 3.76 \Lambda h_{N_2} \right)_j \tag{8.31}$$

$$S_j = \left(\Lambda s_{O_2} + 3.76 \Lambda s_{N_2} \right)_j \tag{8.32}$$

where $j = 1, 2$.

Within the combustor, methane is burned according to the following reaction.

$$CH_4 + \Lambda(O_2 + 3.76N_2) \rightarrow CO_2 + 2H_2O + (\Lambda - 2)O_2 + 3.76\Lambda N_2 \quad (8.33)$$

The thermodynamic model of the fuel compressor (FC) is like that of the air compressor. The combustion products leave the combustor at adiabatic flame temperature T_3, which is the turbine inlet temperature (TIT). The molar flowrate of the air required per unit molar flowrate of the fuel, i.e., 4.76Λ, can be obtained by applying an energy balance to the combustor.

$$h_{CH_4} + \Lambda(h_{2,O_2} + 3.76h_{2,N_2}) = h_{3,CO_2} + 2h_{3,H_2O} + (\Lambda - 2)h_{3,O_2} + 3.76\Lambda h_{3,N_2} \quad (8.34)$$

For a given T_3, Eq. (8.34) is solved to find Λ.

The hot combustion products are expanded within the turbine to the ambient pressure, i.e., $p_4 = p_0$. The temperature at the outlet of the turbine under isentropic condition is found as

$$T_{4s} = T_3 \left(\frac{p_4}{p_3}\right)^{\frac{\gamma_t - 1}{\gamma_t}} \quad (8.35)$$

where $\gamma_t = c_{p,mix}/(c_{p,mix} - R)$ and

$$c_{p,mix} = \sum y_i c_{p,i} \quad i = CO_2, H_2O, O_2, N_2 \quad (8.36)$$

where y_i is the mole fraction of component i in the combustion gases.

The outlet temperature of the turbine T_4 is determined using its isentropic efficiency.

$$\eta_t = \frac{h_3 - h_4}{h_3 - h_{4s}} \quad (8.37)$$

The enthalpy and entropy of the combustion products at the entrance and exit of the turbine are computed by

$$H_j = [h_{CO_2} + 2h_{H_2O} + (\Lambda - 2)h_{O_2} + 3.76\Lambda h_{N_2}]_j \quad (8.38)$$

$$S_j = [s_{CO_2} + 2s_{H_2O} + (\Lambda - 2)s_{O_2} + 3.76\Lambda s_{N_2}]_j \quad (8.39)$$

where $j = 3, 4$.

8.4.2 Determination of SEG

To calculate the specific entropy generation, we note that entropy is generated due to the following four irreversible processes: compression, combustion, expansion, and exhaust cooling. The entropy generation rate per molar flowrate of the fuel within the compressor and turbine is simply obtained from Eqs. (8.40) and (8.41), respectively.

$$\Phi_c = S_2 - S_1 \tag{8.40}$$

$$\Phi_t = S_4 - S_3 \tag{8.41}$$

The rate of entropy generation per unit molar flowrate of the fuel within the combustor is the difference between the entropies of the reactants and the products in reaction (8.33)

$$\Phi_{com} = S_P - S_R \tag{8.42}$$

where $S_P = S_3$, which can be computed using Eq. (8.39), and

$$S_R = s_{CH_4} + \Lambda s_{2,O_2} + 3.76\Lambda s_{2,N_2} \tag{8.43}$$

The last source of entropy generation is the cooling process of the combustion gases that leave the turbine at temperature T_4 and cool down to the ambient temperature. Hence,

$$\Phi_L = \frac{Q_L}{T_0} + (S_5 - S_4) \tag{8.44}$$

where S_4 is determined using Eq. (8.39), and S_5 denotes the entropy of the combustion gases at the ambient temperature and pressure. Hence,

$$S_5 = \left[s_{CO_2} + 2s_{H_2O} + (\Lambda - 2)s_{O_2} + 3.76\Lambda s_{N_2}\right]_{T_0, p_0} \tag{8.45}$$

In Eq. (8.44), Q_L denotes the amount of heat rejected by the cycle per unit flowrate of the fuel.

$$Q_L = HV - W_{net} \tag{8.46}$$

where HV denotes the fuel heating value and W_{net} is the net power produced per unit molar flowrate of the fuel burned in the combustor.

$$W_{net} = W_t - W_c - W_{fc} \tag{8.47}$$

where W_c is the air compressor power consumption, W_t is the turbine power production, and W_{fc} is the power requirement of the fuel compressor.

As the entropy generation terms in Eqs. (8.40)–(8.42) and (8.44) are per unit flowrate of the fuel, the specific entropy generation of the cycle is determined by adding these terms as well as the entropy generation of the fuel compressor, Φ_{fc}. Hence,

$$SEG = \Phi_c + \Phi_t + \Phi_{com} + \Phi_L + \Phi_{fc} \tag{8.48}$$

8.4.3 Illustrative example

Fig. 8.3 depicts illustrative numerical results where the thermal efficiency and SEG are plotted versus the pressure ratio at three different values of the turbine inlet temperature. The main observation is that a higher thermal efficiency corresponds to a lower SEG. The maximum thermal efficiency and the minimum SEG occur at the same pressure ratio for each value of TIT. The pressure ratios at which the thermal efficiency is a maximum and SEG is a minimum are 18, 23.7, and 31 for a TIT of 1173, 1273, and 1373 K, respectively. The respected maximum thermal efficiency is 0.343, 0.370, and 0.395 at the above values of TIT.

The contribution of each irreversible process to the minimum SEG is shown in Fig. 8.4 for three values of TIT. The largest source of the SEG

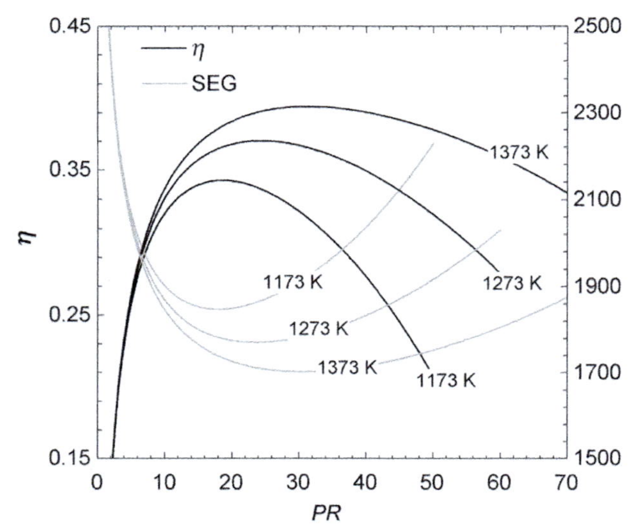

Fig. 8.3 Variation of the thermal efficiency and specific entropy generation of the gas turbine cycle with the pressure ratio at three different values of TIT ($\eta_c = \eta_{fc} = 0.85$, $\eta_t = 0.90$, $T_0 = 298.15$ *K*, $p_0 = 1$ bar).

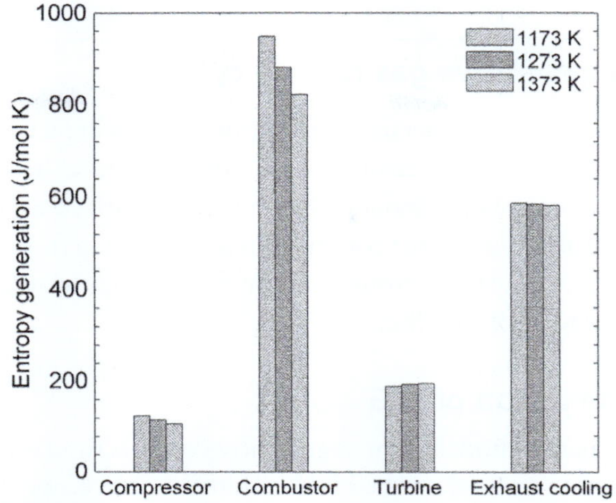

Fig. 8.4 Distribution of the specific entropy generation throughout the gas turbine cycle at the condition of minimum SEG for three values of TIT.

is the combustion process followed by the exhaust gas cooling process. The irreversibility of the fuel compressor is negligible compared to the other components, so it is excluded in Fig. 8.4. Except the turbine, the contribution of all the irreversible processes to the minimum SEG slightly decreases as TIT increases. However, this trend is opposite for the expansion process within the turbine. As seen in Fig. 8.3 that the thermal efficiency improves by an increase in TIT. In practice, the upper limit of TIT is constrained by the durability of the turbine blades.

A further observation in Fig. 8.4 is that a major source of inefficiency in the gas turbine cycle is the significant amount of thermal energy that is wasted with no further use in the downstream. In modern designs of gas turbine engines, a recuperator is placed after the turbine to recover part of the thermal energy of the hot flue gases. Depending on the application, the recovered heat is used to either preheat the pressurized air before the combustor or produce superheated steam to run a downstream Rankine cycle. The former design is referred to as the regenerative gas turbine cycle, and the latter is known as the combined cycle.

Recycling the heat recovered from the hot exhaust gases back to the process boosts the overall power generation efficiency. Both regenerative gas turbine and combined cycles will be discussed in the upcoming sections. Further improvement in the efficiency of the gas turbine cycle would be possible by enhancing the efficiencies of the compressor and turbine.

8.5 Regenerative gas turbine cycle

A schematic of a regenerative gas turbine cycle is depicted in Fig. 8.5. It consists of an air compressor, a fuel compressor, a heat exchanger (recuperator), a combustor, and a turbine. The pressurized air is preheated in the recuperator using the hot combustion gases coming from the turbine. This design reduces the fuel consumption of the cycle compared to the simple gas turbine cycle of Fig. 8.2.

8.5.1 Determination of SEG

The thermodynamic model of the regenerative cycle is similar to that of the simple gas turbine cycle described in Section 8.4. Here, we need to also account for the temperature increase of the air after the compressor by including a recuperator model.

The specific enthalpy of the air at states 2 and 3 in Fig. 8.5 can be related to the specific enthalpy of the flue gases at state 5, h_5, using the effectiveness of the heat exchanger. It is defined as the ratio of the heat transfer from the hot to the cold stream to the maximum possible heat transfer.

$$\epsilon = \frac{Q}{Q_{max}} = \frac{h_3 - h_2}{h_5 - h_2} \tag{8.49}$$

The pressurized air is preheated in the heat exchanger by recovering a portion of the thermal energy of the exhaust flue gases leaving the turbine. Applying the first law to the heat exchanger, one obtains

$$H_3 - H_2 = H_5 - H_6 \tag{8.50}$$

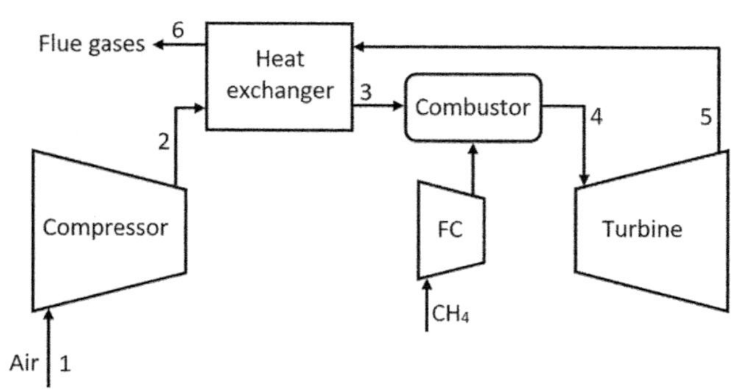

Fig. 8.5 Schematic of a regenerative gas turbine cycle.

The total enthalpy at states 5 and 6 is determined using Eq. (8.38) with $j=5,6$. The temperature of the exhaust gases at state 6 can be obtained using Eq. (8.50).

The entropy generation due to the heat transfer process in the heat exchanger is obtained by

$$\Phi_h = (S_3 - S_2) + (S_6 - S_5) \tag{8.51}$$

where S_5 and S_6 are determined using Eq. (8.39) with $j=5,6$.

Like the simple gas turbine cycle, the flue gases are discharged to the atmosphere. Thus, one would need to account for the entropy production due to the cooling process of the flue gases from temperature T_6 down to the ambient temperature. Hence,

$$\Phi_L = \frac{Q_L}{T_0} + (S_7 - S_6) \tag{8.52}$$

where S_7 is evaluated at the ambient temperature and pressure.

The SEG of the cycle is computed using Eq. (8.53).

$$\text{SEG} = \Phi_c + \Phi_h + \Phi_t + \Phi_{com} + \Phi_L + \Phi_{fc} \tag{8.53}$$

Note that all Φ terms in Eq. (8.53) have the units of $\text{J}/(\text{mol}_{\text{fuel}}\ \text{K})$.

8.5.2 Numerical example

The influence of heat exchanger effectiveness on the maximum thermal efficiency of the gas turbine cycle is presented in Table 8.2 for TIT = 1273 K. The figures in Table 8.2 are obtained using the same parameters given in the caption of Fig. 8.3. The minimum SEG values are also provided in Table 8.2. The results confirm the inverse relation between the thermal efficiency and SEG. An inclusion of a recuperator with an effectiveness of 0.55 may only yield one percentage point increase in the cycle efficiency. A regenerative gas turbine cycle may not be practically advantageous over a simple cycle at low values of the effectiveness.

The advantage of employing a heat exchanger with $\epsilon \geq 0.75$ is evident in Table 8.2. The net increase in the maximum cycle efficiency compared to the simple cycle is 4.5 percentage points at $\epsilon = 0.75$, which raises to 7.8 percentage points at $\epsilon = 0.85$. In the limit of ideal heat exchanger where $\epsilon \rightarrow 1$, the maximum thermal efficiency of the cycle approaches 0.636. It must be noted that the maximum efficiency limit of 0.636 is obtained for

Table 8.2 The effect of heat exchanger effectiveness on the maximum thermal efficiency and minimum specific entropy generation of the regenerative gas turbine cycle.

	Simple cycle	$\epsilon=0.55$	$\epsilon=0.65$	$\epsilon=0.75$	$\epsilon=0.85$	$\epsilon=0.95$
η_{max}	0.370	0.380	0.394	0.415	0.448	0.512
SEG_{min} (J/mol K)	1771	1741	1705	1652	1567	1403

TIT = 1273 K, $\eta_c=0.85$, and $\eta_t=0.90$. So, the efficiency limit would change with a change in TIT, η_c, or η_t.

Fig. 8.6 depicts the distribution of SEG throughout the cycle at the condition of minimum SEG for three different values of ϵ. Like in the simple gas turbine cycle, the combustion process and flue gases cooling are the two major SEG contributors. Evident from Fig. 8.6 is that an improvement in the effectiveness leads to a reduction in the SEG share of both combustion and cooling processes indicating that the thermal energy is utilized more efficiently. On the other hand, the heat exchanger contribution to SEG increases with an increase in the effectiveness, which is due to an increased heat transfer within the recuperator. The combined SEG share of the compressor and turbine remains relatively unaltered. The net reduction in the combined SEG share of the combustion and cooling processes exceeds

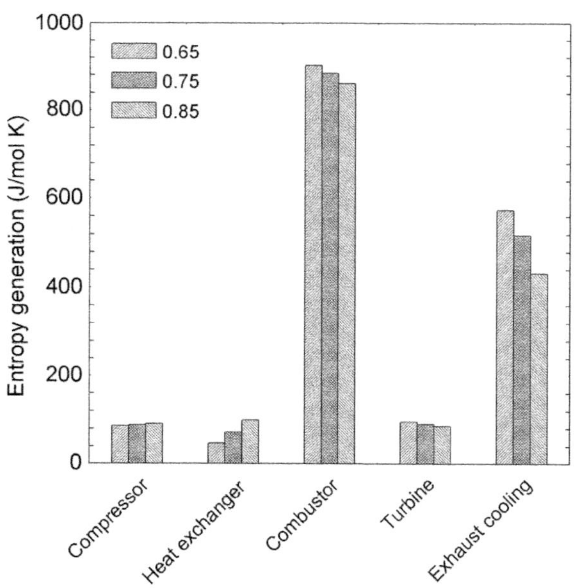

Fig. 8.6 Distribution of the specific entropy generation throughout the regenerative gas turbine cycle at minimum SEG for three values of ϵ ($TIT=1273$ K, $\eta_c=0.85$, $\eta_t=0.90$).

the increased share of the recuperator as ϵ increases. Thus, a higher ϵ leads to a lower SEG and a higher efficiency, which was also noted in Table 8.2.

8.6 Combined cycle

A schematic of a combined cycle is shown in Fig. 8.7. It consists of a gas turbine cycle and a bottoming steam cycle. The hot flue gases leaving the gas turbine (state 4) are directed to a recuperator where the thermal energy of the gases is used to generate steam. For this, the recuperator is referred to as Heat Recovery Steam Generator (HRSG). As shown in Fig. 8.7, the bottoming cycle includes a steam turbine (ST), a condenser (CON), a condensate pump (CP), a deaerator (DEA), and a feed water pump (FP). The pressurized and preheated water (state 9) is superheated to state 10 within the HRSG and the flue gases are discharged to the atmosphere.

To preheat the water before entering the HRSG, steam is extracted from the steam turbine (state 11), which is mixed with water (state 7) in the deaerator. The condensate pump located downstream of the condenser

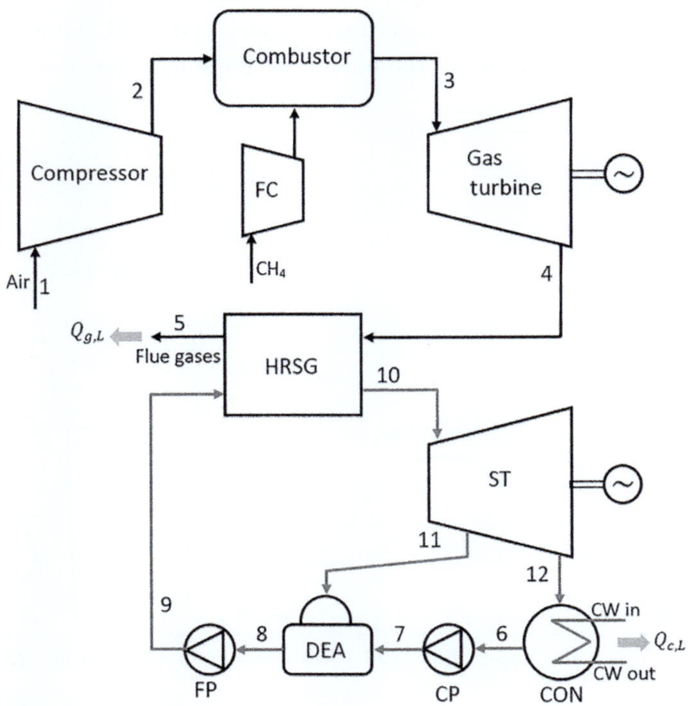

Fig. 8.7 Schematic of a combined cycle.

pressurizes the condensate to the steam extraction pressure, i.e., $p_7 = p_{11}$. The feed water pump further increases the pressure of the preheated water leaving the deaerator to a desired pressure p_9. The saturated steam leaving the steam turbine (state 12) is condensed at a subatmospheric pressure within the condenser. The condensation heat is then transferred to the cooling water (CW), which is cooled down to a desired temperature and recirculated back to the condenser. In fact, the cooling water is a medium between the condenser and the ambient. It first absorbs the condensation heat and then rejects it to the ambient. A *T-s* diagram of the combined cycle is shown in Fig. 8.8 for a TIT of 1273 K.

8.6.1 Thermodynamic model

For simplicity of the analysis, the pressure drop is neglected on the path of the working fluid. The thermodynamic modeling is only focused on the components of the steam cycle since we have previously discussed the thermodynamic model of the gas turbine cycle. Table 8.3 provides the equations of the conservation of energy and the entropy generation rate per unit molar flowrate of the fuel for the components of the steam cycle. In Table 8.3, n_{st} denotes the molar flowrate of the steam produced in the HRSG, whereas n_{cw} denotes the molar flowrate of the condenser cooling water, both per unit

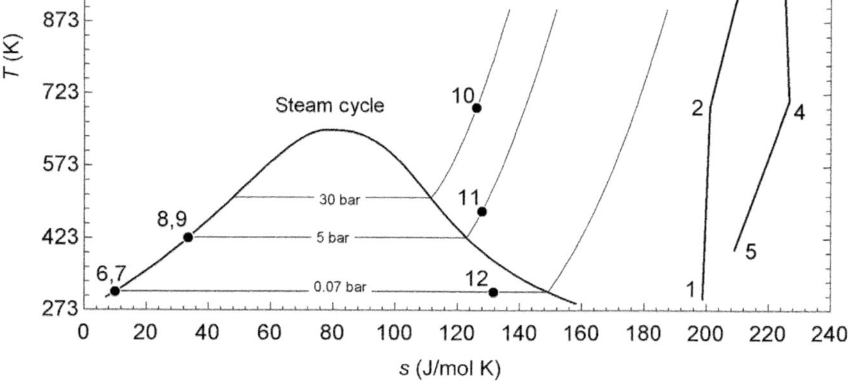

Fig. 8.8 *T-s* diagram of the combined cycle at TIT = 1273 K.

Table 8.3 The energy conservation and entropy balance equations of the steam cycle components.

Component	Energy conservation equation	Entropy generation equation
HRSG	$H_4 - H_5 = n_{st}(h_{10} - h_9)$[a]	$\Phi_{HRSG} = (S_5 - S_4) + n_{st}(s_{10} - s_9)$
Steam turbine	$W_{st} = n_{st}h_{10} - n_{11}h_{11} - n_{12}h_{12}$	$\Phi_{st} = n_{11}s_{11} + n_{12}s_{12} - n_{st}s_{10}$
Deaerator	$n_{11}h_{11} + n_7h_7 = n_8h_8$	$\Phi_{dea} = n_8s_8 - n_7s_7 - n_{11}s_{11}$
Condenser	$n_6(h_{12} - h_6) = n_{cw}(h_{cw,\ out} - h_{cw,\ in})$	$\Phi_{con} = n_6(s_6 - s_{12}) + n_{cw}(s_{cw,\ out} - s_{cw,\ in})$
Condensate pump	$W_{cp} = n_6v_6(p_7 - p_6)/\eta_{cp}$	$\Phi_{cp} = n_6(s_7 - s_6)$
Feed pump	$W_{fp} = n_{st}v_8(p_9 - p_8)/\eta_{cp}$	$\Phi_{fp} = n_{st}(s_9 - s_8)$

[a] All molar flowrates are per unit molar flowrate of the fuel.

molar flowrate of the fuel. Also, the isentropic efficiencies of the condensate and feed water pumps are denoted by η_{cp} and η_{fp}, respectively. Note that the mass conservation requires $n_8 = n_9 = n_{10}$, $n_6 = n_7 = n_{12}$, and $n_{10} = n_{11} + n_{12}$.

The thermal efficiency of the combined cycle is calculated by $\eta = \frac{W_{net,cc}}{HV}$. The net power production of the combined cycle per unit flowrate of the fuel is determined as follows.

$$W_{net,cc} = \left(W_t - W_c - W_{fc}\right) + \left(W_{st} - W_{cp} - W_{fp}\right) \tag{8.54}$$

To accurately compute the total entropy generation of the combined cycle, we need to account for the irreversible processes due to the heat rejection from the cycle to the ambient through the flue gases and the cooling water, to be denoted by $Q_{g,L}$ and $Q_{c,L}$, respectively; see Fig. 8.7.

$$Q_{c,L} = n_6(h_{12} - h_6) \tag{8.55}$$

$$Q_{g,L} = HV - W_{net,cc} - Q_{c,L} \tag{8.56}$$

The entropy generation due to discharging the condensation heat by the cooling water to the environment is obtained by

$$\Phi_{c,L} = \frac{Q_{c,L}}{T_0} + n_{cw}\left(s_{cw,in} - s_{cw,out}\right) \tag{8.57}$$

Likewise, rejecting the heat of the flue gases to the environment generates entropy. Hence,

$$\Phi_{g,L} = \frac{Q_{g,L}}{T_0} + (S_{13} - S_5) \tag{8.58}$$

where S_{13} is the entropy of the flue gases at the ambient pressure and temperature.

Finally, the specific entropy generation of the cycle is determined by adding the entropy generation within the gas cycle (GC), steam cycle (SC), and due to rejecting waste heat to the environment.

$$\text{SEG} = \sum_{GC} \Phi_i + \sum_{SC} \Phi_j + \left(\Phi_{g,L} + \Phi_{c,L}\right) \tag{8.59}$$

where i and j refer to the components of the topping and bottoming cycles, respectively.

8.6.2 Illustrative example

The thermal efficiency and specific entropy generation of the combined cycle are plotted against the compressor pressure ratio in Fig. 8.9 for three values of TIT. These results are obtained for the operating parameters given in Table 8.4. The steam temperature is set to be 15 K below the temperature of the combustion gases entering the HRSG [3]. The thermal efficiency peaks at a pressure ratio at which the specific entropy generation approaches a minimum value. Fig. 8.9 shows that a design of the combined cycle power plant based on the maximum thermal efficiency is the same as that of the minimum specific entropy generation.

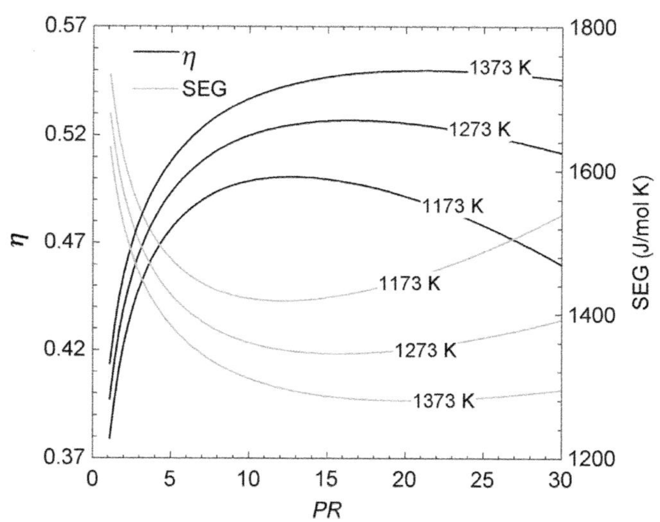

Fig. 8.9 Variation of the thermal efficiency and specific entropy generation of the combined cycle with the pressure ratio at three different values of TIT.

Table 8.4 The operating parameters of the combined cycle.

Parameter	Quantity
Gas cycle	
Ambient temperature, T_0	298.15 K
Ambient pressure, p_0	1 bar
Gas turbine inlet temperature, T_3	1273 K
Compressor isentropic efficiency η_c	0.85
Gas turbine isentropic efficiency, η_t	0.90
Steam cycle	
Steam turbine inlet pressure, p_{10}	30 bar
Condenser pressure, p_6	0.07 bar
Exhaust temperature of flue gases, T_5	400 K
Steam turbine isentropic efficiency, η_{st}	0.90
Condensate pump isentropic efficiency, η_{cp}	0.70
Feed pump isentropic efficiency, η_{fp}	0.70

The contribution of the topping gas cycle, bottoming steam cycle, and waste heat rejection to the minimum SEG of the combined cycle is illustrated in Fig. 8.10 at three different values of TIT. The share of the gas turbine cycle to the inefficiencies of the combined cycle is five to six times greater than the combined share of the steam cycle and waste heat rejection. Thus, in order to improve the efficiency of the combined cycle,

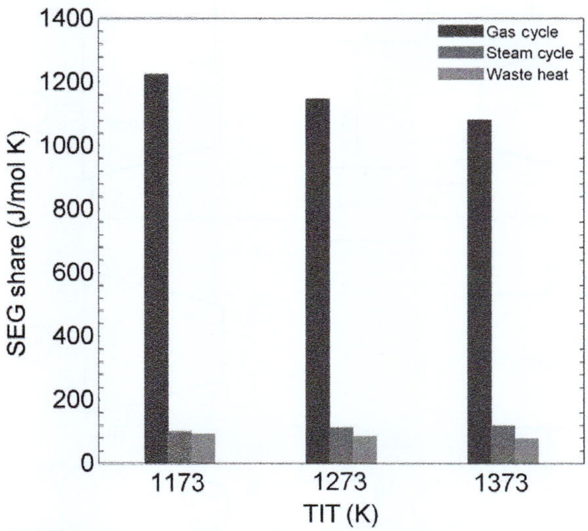

Fig. 8.10 Contribution of the gas cycle, steam cycle, and waste heat to the minimum SEG at three values of TIT.

one must focus on enhancing the design of the topping cycle. For this, a designer should look for ways to utilize the thermal energy efficiently. The design of the combined cycle may be altered by including a heat exchanger between the gas turbine and the HRSG to preheat the air before entering the combustor; see Fig. 8.11. The thermal energy of the flue gases would then be recovered within both HRSG and heat exchanger.

8.6.3 Modified design

The modified combined cycle shown in Fig. 8.11 is indeed an integrated regenerative gas turbine and a steam cycle. To examine the performance of the proposed layout, the cycle efficiency is calculated using the same data given in Table 8.4 and plotted against the compressor pressure ratio in Fig. 8.12 (denoted by HRSG &H.Ex) for TIT = 1273 K and $\epsilon = 0.85$. The thermal efficiency of the combined cycle of Fig. 8.7 (denoted by HRSG Only) is also provided in Fig. 8.12 for comparison.

Two main observations can be made in Fig. 8.12. First, the maximum thermal efficiency of the cycle with "HRSG & H.Ex" is 0.564, whereas that of the combined cycle with "HRSG Only" is 0.527. Second, the maximum

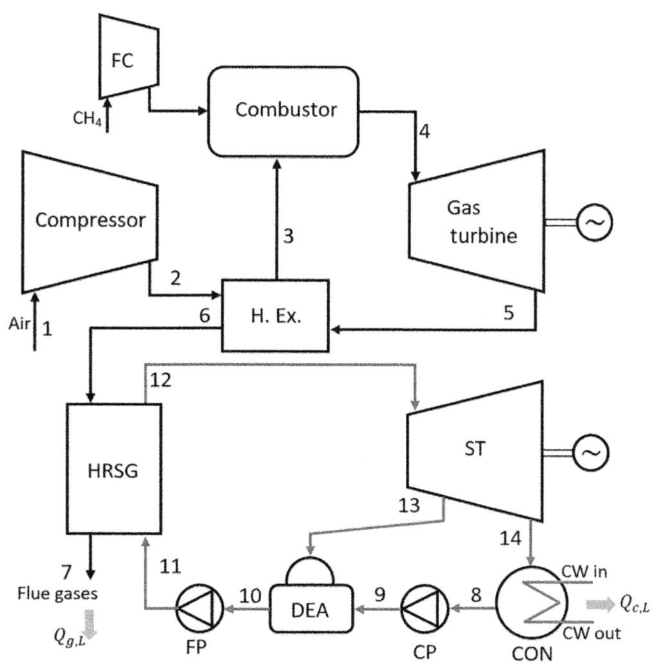

Fig. 8.11 The modified layout of the combined gas/steam turbine power cycle.

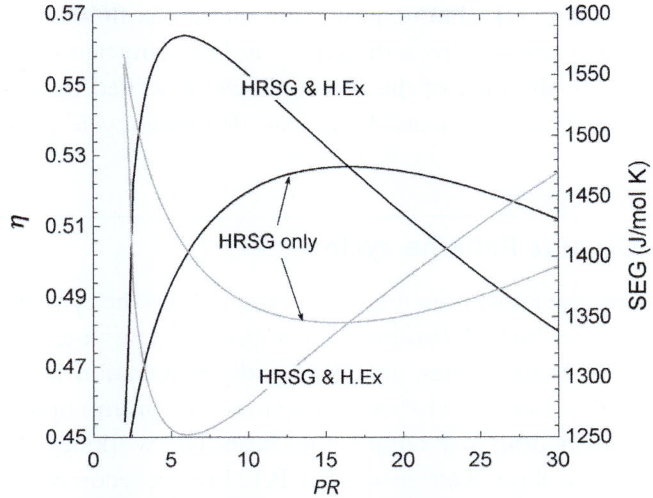

Fig. 8.12 Comparison of the efficiency and SEG of the combined cycles of Figs. 8.7 and 8.11 at TIT = 1273 K.

efficiency takes place at an optimum pressure ratio of 5.7 in the former and 15.9 in the latter. So, the combined regenerative gas turbine and steam cycle is expected to be 3.7 percentage points more efficient with smaller turbomachinery in the topping cycle compared to the cycle of Fig. 8.7.

The performances of the two designs of the combined cycle at maximum efficiency are compared in Table 8.5. The results suggest that recovering the thermal energy of the exhaust gases is more efficient in the modified cycle (Fig. 8.11) than the original design (Fig. 8.7). The modified cycle is expected

Table 8.5 Performance comparison of the combined cycles of Figs. 8.7 and 8.11.

Parameter	Combined cycle	Modified cycle
Optimum pressure ratio	15.9	5.7
HRSG inlet temperature (K)	709.6	587.3
Steam temperature (K)	694.6	572.3
Minimum SEG (J/K mole fuel)	1346	1253
Efficiency of combined cycle	0.527	0.564
Efficiency of topping cycle	0.360	0.445
Efficiency of bottoming cycle	0.343	0.328
Power output (kW)[a]	422.8	452.4
Power output of topping cycle (kW)[a]	289.1	356.9
Power output of bottoming cycle (kW)[a]	133.7	95.5
Steam flowrate (mol/s)[a]	8.20	6.88

[a] Per unit molar flowrate of the fuel.

to produce roughly 30 kW more power per unit molar flowrate of the fuel than the original design. Given the relatively low temperature of the flue gases (587.3 K) at the inlet of the HRSG in the modified cycle, it would rather be more practical to use an organic Rankine cycle (ORC) than the steam cycle.

8.7 Organic Rankine cycle

The basic operating principles of an organic Rankine cycle (ORC) are similar to the steam cycle. It consists of an evaporator, a turbine, a condenser, and a pump. In some studies, an ORC with regeneration has also been examined. ORCs have widely been recognized as a promising technology to produce power from low-temperature heat. The working substance of ORC is an organic fluid. Refrigerants like R1234yf are recommended fluids for use in ORCs for low-to-moderate temperature ($\leq$473 K) heat sources [4]. At relatively high heat source temperatures ($\sim$573 K), fluids with high critical temperatures like toluene are employed [5]. Table 8.6 lists the properties of some working fluids employed in ORCs.

The common applications of ORCs include geothermal power plants, biomass combined heat and power plants, integrated solar and ORC systems, industrial waste heat recovery, integrated internal combustion engine and ORC systems [6]. The combined regenerative gas turbine and ORC has also been reported. The flue gases temperature is relatively low at the exhaust of regenerative gas turbines, e.g., 587.3 K in Table 8.5. So, it would be impractical to operate traditional combined gas/steam cycle with a low exhaust temperature at high steam pressures mainly due to the significant

Table 8.6 Physical and environmental impact properties of common organic fluids.

Fluid	MW (g/mol)	Formula	T_c (K)[a]	p_c (bar)[b]	T_b (K)[c]	ODP[d]	GWP[e]
R123	152.9	$C_2HF_3Cl_2$	456.8	36.68	300.9	0.02	77
R1234yf	114.0	$C_3H_2F_4$	367.9	33.82	243.7	0	4
R134a	102.0	$C_2H_2F_4$	374.2	40.59	247.1	0	1430
R152a	66.05	$C_2H_4F_2$	386.4	45.20	249.1	0	124
R245fa	134.0	$C_3H_3F_5$	427.2	36.51	288.2	0	1030
R600	58.12	C_4H_{10}	425.1	37.96	272.6	0	4
R600a	58.12	C_4H_{10}	407.8	36.40	261.5	0	3

[a] Critical temperature.
[b] Critical pressure.
[c] Boiling temperature.
[d] Ozone depletion potential.
[e] Global warming potential (100-year).

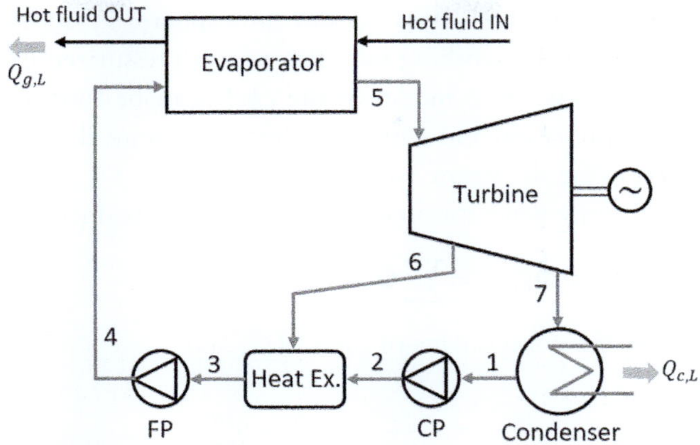

Fig. 8.13 Schematic of an organic Rankine cycle.

moisture of the saturated vapor at the last stage of steam turbine. Integration of a bottoming ORC with regenerative gas turbines or gas turbines with a relatively high pressure ratio is practically viable [5].

A schematic of an ORC with regeneration is shown in Fig. 8.13. A *T-s* diagram of the cycle operating with R600 as the working fluid is depicted in Fig. 8.14. A hot fluid with a given temperature and flowrate enters the evaporator where the preheated working fluid is further heated up to a few degrees above its evaporation temperature. The hot fluid with

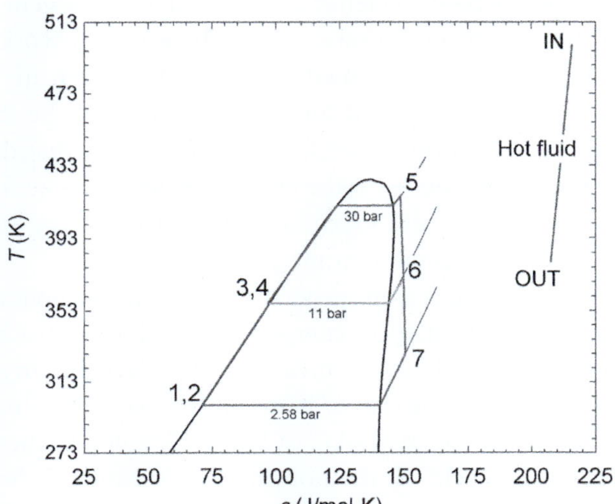

Fig. 8.14 *T-s* diagram of the organic Rankine cycle operating with R600.

a low temperature is discharged to the atmosphere. A regenerator is employed to preheat the working fluid before it is pressurized by the feed pump. The thermodynamic model of the cycle can be built up using a set of energy conservation and entropy balance equations described in the preceding section for the steam cycle.

The total entropy generation rate of the ORC is obtained by

$$\dot{\Phi}_{\text{ORC}} = \sum \dot{\Phi}_i + \left(\dot{\Phi}_{g,L} + \dot{\Phi}_{c,L}\right) \tag{8.60}$$

where i denotes the components of the ORC, and

$$\dot{\Phi}_{g,L} = \frac{\dot{Q}_{g,L}}{T_0} + \left(S_0^{HF} - S_{out}^{HF}\right) \tag{8.61}$$

where S_0^{HF} denotes the entropy flow of the hot fluid at the ambient temperature and pressure, and S_{out}^{HF} is its entropy flow at the outlet of the evaporator. The amount of heat discharged by the hot fluid to the environment is determined using Eq. (8.62).

$$\dot{Q}_{g,L} = H_{out}^{HF} - H_0^{HF} \tag{8.62}$$

The thermal efficiency of the cycle is evaluated as follows.

$$\eta = \frac{\dot{W}_{net}}{H_{in}^{HF} - H_{out}^{HF}} \tag{8.63}$$

Fig. 8.15 shows the thermal efficiency and the total entropy generation rate varying with the extraction pressure, p_6, for three typical working fluids. The results in Fig. 8.15 are obtained using the data given in Table 8.7. The efficiencies of the pumps and turbine are assumed to be the same as those in Table 8.4. The main observation in Fig. 8.15 is that the thermal efficiency inversely correlates with the entropy generation rate for all three fluids. A design based on the maximum thermal efficiency is therefore equivalent to that of minimum entropy generation rate.

Notice that the specific entropy generation method is inapplicable for the case of ORC, because the thermal energy is supplied through a hot stream, unlike the cycles examined in the previous sections that are driven by fuel combustion. The inverse correlation between the cycle efficiency and the entropy generation rate for the ORC of Fig. 8.13 is due to the fixed heat input supplied to the cycle via the hot fluid; see Chapter 7, Section 7.7. The reader may also return to Chapter 6, Section 6.5 for the analysis of the Carnot vapor cycle with fixed heat input.

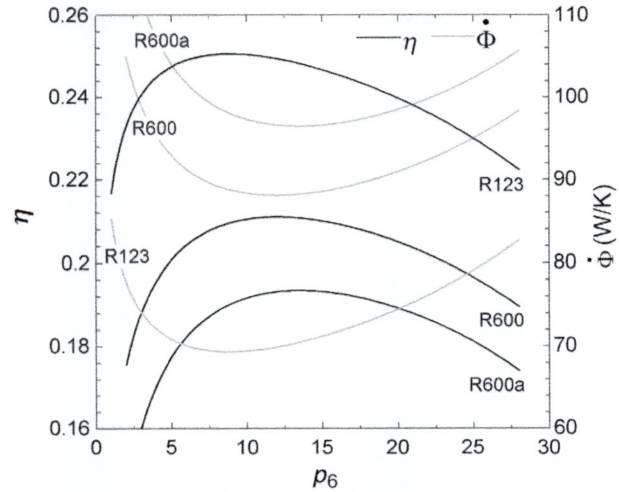

Fig. 8.15 Variation of the thermal efficiency and entropy generation rate of the ORC with the extraction pressure for three different working fluids.

Table 8.7 The ORC data used in the calculations.

Parameter	**Quantity**
Hot fluid inlet temperature (K)	500
Hot fluid outlet temperature (K)	380
Turbine inlet pressure (bar)	30
Turbine inlet temperature (K)	$T_{evp}+5$
Condensation temperature (K)	300

References

[1] Y. Cengel, M. A. Boles, Thermodynamics: An Engineering Approach, fifth ed., McGraw-Hills, New York, Chapter 15.

[2] P.J. Linstrom, W.G. Mallard, NIST Chemistry WebBook, NIST Standard Reference Database Number 69, National Institute of Standards and Technology, Gaithersburg, MD, 2017. http://webbook.nist.gov/.

[3] I.S. Ertesvåg, H.M. Kvamsdal, O. Bolland, Exergy analysis of a gas-turbine combined-cycle power plant with precombustion CO_2 capture, Energy 30 (2005) 5–39.

[4] S. Quoilin, S. Declaye, B.F. Tchanche, V. Lemort, Thermo-economic optimization of waste heat recovery organic Rankine cycles, Appl. Therm. Eng. 31 (2011) 2885–2893.

[5] R. Chacartegui, D. Sánchez, J.M. Muñoz, T. Sánchez, Alternative ORC bottoming cycles FOR combined cycle power plants, Appl. Energy 86 (2009) 2162–2170.

[6] S. Quoilin, M.V.D. Broek, S. Declaye, P. Dewallef, V. Lemort, Techno-economic survey of organic Rankine cycle (ORC) systems, Renew. Sust. Energy Rev. 22 (2013) 168–186.

CHAPTER NINE

Entropy and fuel cells

9.1 Introduction

The power-generating systems that we have discussed so far are heat engines, which produce useful work by converting a portion of heat supplied through a high-temperature reservoir, fuel combustion, or a hot stream. Fuel cell is a device that produces electrical energy by converting the chemical energy of a fuel through a series of electrochemical reactions. A fuel cell consists of the following main segments: anode, cathode, and electrolyte (Fig. 9.1) The primary fuel of most fuel cells is hydrogen, which is supplied to the anode. The cathode is fed by an oxidizer, i.e., oxygen or air. The reactions taking place in the anode and cathode of a fuel cell include

$$\text{Anode}: H_2 + O^{=} \rightarrow H_2O + 2e^{-} \tag{9.1}$$

$$\text{Cathode}: \frac{1}{2}O_2 + 2e^{-} \rightarrow O^{=} \tag{9.2}$$

The electrochemical reaction taking place in the anode between the hydrogen and oxygen ions leads to water production and a release of electrons. In the cathode, oxygen reacts with the electrons, which are transferred from the anode through an external circuit. On the other hand, the oxygen ions produced in the cathode travel across the electrolyte to the anode. The net cell reaction can be expressed as

$$H_2 + \frac{1}{2}O_2 \rightarrow H_2O \tag{9.3}$$

The byproducts of the cell reaction are water, electricity, and heat generated due to various losses, including activation polarization, ohmic, and concentration losses. Further details about the operation of various types of fuel cell can be found in Refs. [1, 2].

Fuel cells have been regarded as highly efficient devices with applications in transportation and power sectors. High-temperature fuel cells such as solid oxide fuel cells (SOFC) are excellent candidate for stationary power

Entropy Analysis in Thermal Engineering Systems
https://doi.org/10.1016/B978-0-12-819168-2.00009-X

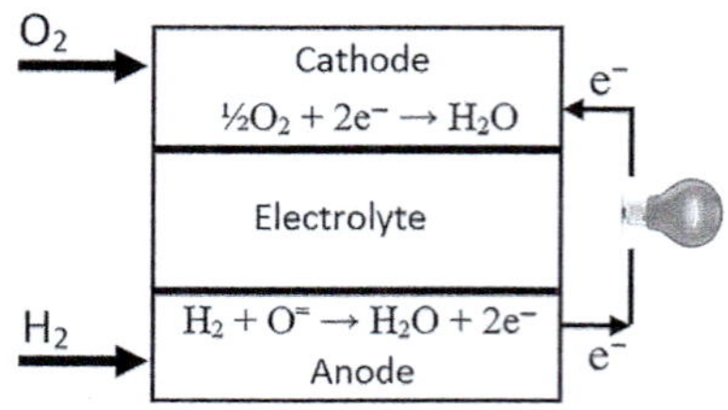

Fig. 9.1 The basic operating principles and components of a fuel cell.

Table 9.1 Comparison of the measured NO_x and CO emissions (g/MWh) from fuel cells and conventional technologies [3].

	Fuel cell	Condensing boiler	CHP engine
NO_x	1–4	58	30–270
CO	1–8	43	10–50

generation where the hot reaction products can be used in a bottoming cycle such as a gas turbine cycle, a steam cycle, or a combined gas/steam cycle to produce additional power. Compared to combustion-based technologies, the operation of fuel cells is environment friendly; a 0.7–1 kW system may have an annual CO_2 saving of 1.3–1.9 tones [3]. Also, the nitride oxides and carbon monoxide emissions from fuel cells are notably less than the conventional systems; see Table 9.1.

Understanding the limit of fuel-to-power conversion efficiency is a fundamental step toward enhancing and optimizing the performance of fuel cell-driven power generating systems. There have been controversial statements in the scientific literature over determination of the maximum efficiency of fuel cells and whether their performance is limited to the Carnot efficiency. Different expressions have been proposed for maximum fuel cell work. Our primary goal is to show how the fundamental thermodynamic laws should be applied to establish an accurate expression for the theoretical maximum work of a hydrogen-fed fuel cell.

9.2 Maximum conversion efficiency

Consider a hydrogen fuel cell (Fig. 9.2) in which oxygen and hydrogen are fed to the cathode and anode, respectively. The operation of the fuel cell is assumed to be steady state, adiabatic, and isobaric. The reactants enter the fuel cell at temperature T_R and pressure p_0. The reaction product (water)

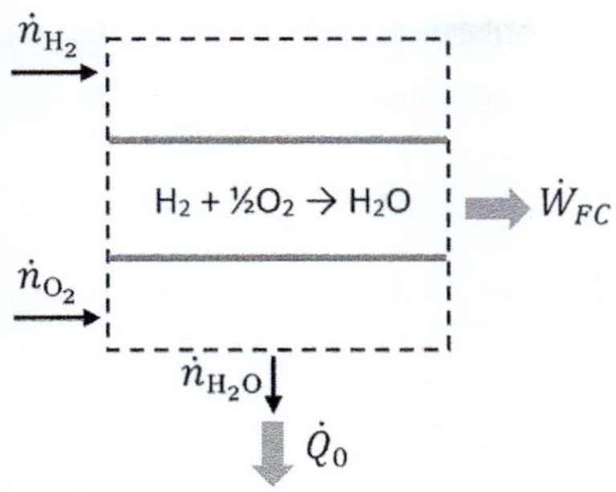

Fig. 9.2 The system boundary of a hydrogen fuel cell.

leaving the cell at temperature T_P is discharged to the surrounding characterized by temperature T_0 and pressure p_0. Applying an energy balance to the fuel cell shown in Fig. 9.2 (the dashed rectangle designates the system boundary), one obtains

$$(\dot{n}_{H_2}h_{H_2} + \dot{n}_{O_2}h_{O_2})_R = \dot{W}_{FC} + (\dot{n}_{H_2O}h_{H_2O})_P \tag{9.4}$$

where $\dot{n}$ is the molar flowrate, h the specific enthalpy, and $\dot{W}_{FC}$ the electricity production of the fuel cell. Note that the enthalpies of hydrogen and oxygen are determined at T_R whereas the water enthalpy is evaluated at T_P.

The entropy generation rate due to the operation of the fuel cell is

$$\dot{\Phi}_{FC} = \dot{n}_{H_2O}(s_{H_2O})_P - (\dot{n}_{H_2}s_{H_2} + \dot{n}_{O_2}s_{O_2})_R \tag{9.5}$$

where the specific entropies are evaluated at pressure p_0.

As shown in Fig. 9.2, water at temperature T_P is discharged to the surrounding. Consistent with our analysis in Chapter 8, the entropy generation due to the cooling process of the reaction product needs to be also accounted for. Thus, the total entropy generation rate associated with the operation of the fuel cell is

$$\dot{\Phi} = \langle\dot{\Phi}\rangle_{FC} + \langle\dot{\Phi}\rangle_{Cooling} \tag{9.6}$$

where

$$\dot{\Phi}_{Cooling} = \frac{\dot{Q}_0}{T_0} + \dot{n}_{H_2O}\left[(s_{H_2O})_0 - (s_{H_2O})_P\right] \tag{9.7}$$

$$\dot{Q}_0 = \dot{n}_{H_2O}\left[(h_{H_2O})_P - (h_{H_2O})_0\right] \tag{9.8}$$

Substituting Eqs. (9.5), (9.7), and (9.8) into Eq. (9.6) yields

$$\dot{\Phi} = \dot{n}_{H_2O}(s_{H_2O})_0 - (\dot{n}_{H_2}s_{H_2} + \dot{n}_{O_2}s_{O_2})_R + \frac{\dot{n}_{H_2O}\left[(h_{H_2O})_P - (h_{H_2O})_0\right]}{T_0} \quad (9.9)$$

Eliminating $(h_{H_2O})_P$ between Eqs. (9.4) and (9.9), we find

$$\dot{\Phi} = \frac{(\dot{n}_{H_2}h_{H_2} + \dot{n}_{O_2}h_{O_2})_R - \dot{n}_{H_2O}(h_{H_2O})_0}{T_0} + \left[\dot{n}_{H_2O}(s_{H_2O})_0 - (\dot{n}_{H_2}s_{H_2} + \dot{n}_{O_2}s_{O_2})_R\right] - \frac{\dot{W}_{FC}}{T_0} \quad (9.10)$$

Setting $\dot{\Phi}$ equal to zero leads to an expression for the maximum fuel cell power. Hence,

$$\left(\dot{W}_{FC}\right)_{max} = \dot{E}_{in} + T_0\left[\dot{n}_{H_2O}(s_{H_2O})_0 - (\dot{n}_{H_2}s_{H_2} + \dot{n}_{O_2}s_{O_2})_R\right] \quad (9.11)$$

where

$$\dot{E}_{in} = (\dot{n}_{H_2}h_{H_2} + \dot{n}_{O_2}\mathrm{h}_{O_2})_R - \dot{n}_{H_2O}(h_{H_2O})_0 \quad (9.12)$$

The expression that we found for the maximum fuel cell power is neither the change in Gibbs function at reaction temperature nor that at the surrounding temperature T_0. However, if the reactants enter the fuel cell at the surrounding temperature, i.e., $T_R = T_0$, Eq. (9.11) reduces to

$$\left(\dot{W}_{FC}\right)_{max} = (G_{H_2} + G_{O_2} - G_{H_2O})_0 \quad (9.13)$$

where G is the Gibbs function defined as

$$G = H - T \cdot S \quad (9.14)$$

In Eq. (9.13), G_i (i: H_2, O_2, H_2O) is calculated at the surrounding temperature and pressure.

The maximum conversion efficiency of the fuel cell is obtained by

$$\eta_{max} = \frac{\left(\dot{W}_{FC}\right)_{max}}{\dot{E}_{in}} = 1 + T_0\frac{\dot{n}_{H_2O}(s_{H_2O})_0 - (\dot{n}_{H_2}s_{H_2} + \dot{n}_{O_2}s_{O_2})_R}{\dot{E}_{in}} \quad (9.15a)$$

Or, per molar flowrate of hydrogen, one may write

$$\eta_{max} = 1 - T_0\frac{(s_{H_2} + 0.5s_{O_2})_R - (s_{H_2O})_0}{(h_{H_2} + 0.5h_{O_2})_R - (h_{H_2O})_0} \quad (9.15b)$$

Also, the amount of heat dissipated to the surrounding at maximum efficiency is

$$\dot{Q}_0 = -T_0\left[\dot{n}_{H_2O}(s_{H_2O})_0 - (\dot{n}_{H_2}s_{H_2} + \dot{n}_{O_2}s_{O_2})_R\right] \quad (9.16)$$

9.2.1 Hydrogen-air fuel cell

A more realistic analysis should account for the fact that oxygen is not freely available in the nature. We consider a hydrogen fuel cell, which uses air, comprising 21% oxygen and 79% nitrogen (mole basis) as the oxidizer. Thus, the overall cell reaction obeys

$$H_2 + \Lambda(O_2 + 3.76N_2) \rightarrow H_2O + (\Lambda - 0.5)O_2 + 3.76\Lambda N_2 \qquad (9.17)$$

where Λ (≥ 0.5) is a stoichiometric coefficient.

As the products cool down to the surrounding temperature, a portion of water vapor may condense and the rest is left as vapor in the products mixture. The derivation procedure for the maximum fuel cell power is the same as described in the previous section. The following expression is obtained for $\left(\dot{W}_{FC}\right)_{max}$.

$$\left(\dot{W}_{FC}\right)_{max} = H_R - H_{P,0} + T_0\left(S_{P,0} - S_R\right) \qquad (9.18)$$

where the reactants include hydrogen, oxygen, and nitrogen, and the products comprise water (liquid and gas), unused oxygen, and nitrogen.

9.2.2 Fuel cell operating on methane

Like oxygen, hydrogen is not freely available in the nature. One would then need to account for production of hydrogen from a primary fuel. For example, hydrogen can be produced through steam reforming of a hydrocarbon fuel. Assume that methane is the fuel that is reformed with steam to produce carbon monoxide and hydrogen.

$$CH_4 + H_2O \rightarrow 3H_2 + CO \qquad (9.19)$$

Additional hydrogen may be produced by converting carbon monoxide to carbon dioxide through water-gas shift reaction.

$$CO + H_2O \rightarrow H_2 + CO_2 \qquad (9.20)$$

The reforming reaction, Eq. (9.19), is endothermic whereas the water-gas shift reaction is exothermic. It can be deduced from Eqs. (9.19) and (9.20) that for each mole of methane, the net reforming plus shift reaction requires two moles of steam to produce four moles of hydrogen.

$$CH_4 + 2H_2O \rightarrow 4H_2 + CO_2 \qquad (9.21)$$

where the overall reaction is endothermic with a reaction enthalpy of 165 kJ/mol at 298 K.

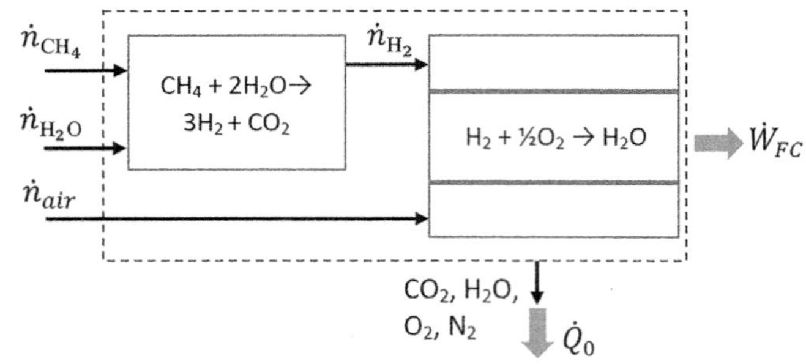

Fig. 9.3 Schematic of a fuel cell and a methane reformer.

The hydrogen produced from methane reforming is fed to the anode to electrochemically react with the oxygen of the air that is supplied to the cathode. So, the system takes in methane, water, and air through three different streams (Fig. 9.3), and exhausts a mixture of carbon dioxide, steam, oxygen, and nitrogen to the environment. The overall chemical reaction of the combined system of reformer and fuel cell may be expressed as

$$CH_4 + 2H_2O + \Lambda(O_2 + 3.76N_2) \rightarrow CO_2 + 4H_2O + (\Lambda - 2)O_2 + 3.76\Lambda N_2 \quad (9.22)$$

The mixture of products leaving the fuel cell at T_P cools down to T_0. The possibility of partial condensation of steam should be accounted for. The general form of the expression for the maximum power production would be the same as that in Eq. (9.18).

9.2.3 Numerical example

The maximum conversion efficiencies of the hydrogen-oxygen, hydrogen-air, and methane-air fuel cells are compared in Fig. 9.4. The trend of the maximum efficiency varying with the reactants temperature is almost the same for all three fuel cells. The maximum conversion efficiency takes place at the reactants temperature of 298.15 K. However, by increasing the reactants temperature, the efficiency first decreases, then remains unaltered over a temperature range, and finally begins to rise. The efficiency plateau for the hydrogen-oxygen, hydrogen-air, and methane-air fuel cells takes place in the range 1350–1540 K, 1170–1300 K, and 1590–1730 K, respectively. At a given reactants temperature, the maximum efficiency of a fuel cell operating on methane-air is the highest, whereas that running on hydrogen-air is the lowest, and that of the hydrogen-oxygen fuel cell in between. The lowest value of the maximum efficiency is 82.1%, 75.7%, and 79.3%,

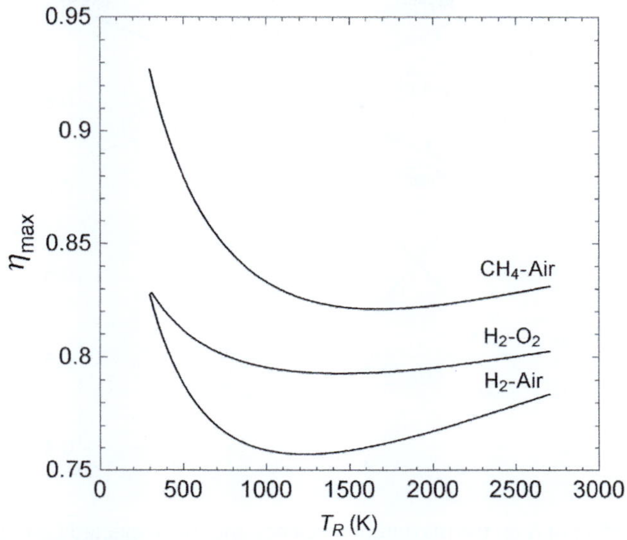

Fig. 9.4 Maximum efficiency comparison at various reactant temperatures ($\Lambda=0.5$ for H_2-air fuel cell, $\Lambda=2$ for CH_4-air fuel cell).

respectively, for CH_4-air, H_2-O_2, and H_2-air fuel cells and the respected highest maximum efficiency is 92.7%, 82.7%, and 82.7%.

A further comparison is shown in Fig. 9.5 where the effect of Λ is examined on the maximum efficiencies of the hydrogen-air and methane-air fuel cells, where in the former $\Lambda \geq 0.5$, and in the latter $\Lambda \geq 2$. In both devices, the maximum efficiency increases linearly with the stoichiometric coefficient with the slope of $\eta_{max} - \Lambda$ being greater for the hydrogen-air fuel cell than the methane-air fuel cell. As seen in Fig. 9.5, the maximum efficiency of 100% is achieved at $\Lambda=7.2$ and $\Lambda=9.8$ in H_2-air and CH_4-air fuel cells, respectively. An efficiency of 100% corresponds to a case where the entire energy supplied to the fuel cell is converted into electrical energy, and the entropy difference between the products at T_0, p_0 and reactants at the inlet of the fuel cell is zero.

9.3 Open circuit voltage

The maximum fuel cell power can be used to determine the cell voltage at the reversible limit. The relation between the fuel cell work and voltage obeys

$$W_{FC} = N_e F V \tag{9.23}$$

where N_e denotes the number of electrons, and F is the Faraday constant (= 96,485C).

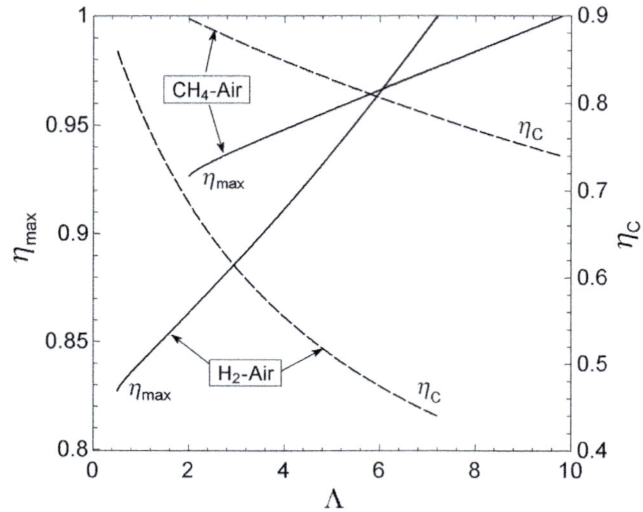

Fig. 9.5 The effect of Λ on the maximum efficiency and the respected Carnot efficiency, $T_R = 298$ K.

The open circuit voltage corresponds to the operation of the fuel cell at the reversible limit. So, with the use of Eq. (9.23), we get

$$V_{OC} = \frac{(W_{FC})_{max}}{N_e F} \tag{9.24}$$

For example, the theoretical maximum work of H_2-O_2 fuel cell at $T_R = 298.15$ K is calculated to be 237.1 kJ/mol of hydrogen assuming surrounding temperature and pressure of 298.15 K and 1 bar. Also, the number of electrons per each mole of hydrogen traveling from the anode to the cathode is 2. Substituting these values into Eq. (9.24), we find

$$V_{OC} = \frac{(W_{FC})_{max}}{N_e F} = \frac{237.1 \times 10^3}{(2)(96,485)} = 1.229\ (\text{V})$$

Table 9.2 provides the open circuit voltage of the fuel cells discussed in the preceding section at three different reactants temperatures.

Table 9.2 Open circuit voltage (measured in volts) calculated at three different reactants temperatures ($T_0 = 298.15$ K, $p_0 = 1$ bar).

T_R (K)	H_2-O_2	H_2-Air[a]	CH_4-Air[b]
298	1.229	1.214	1.055
373	1.231	1.218	1.058
1273	1.361	1.514	1.317

[a] $\Lambda = 0.5$.
[b] $\Lambda = 2.0$.

9.4 Misconceptions

Two misconceptions with respect to the operation of fuel cells need to be carefully addressed. The first issue is the incorrect claim(s) about the possibility of fuel cell efficiency being >100%, when the entropy of the reaction products is greater than that of the reactants. This misconception has been addressed to some extent by others, e.g., Ref. [4]. The second misconception is with respect to inaccurate calculations, which have led to a claim that the upper bound of fuel cell efficiency is limited to the Carnot efficiency.

9.4.1 First issue: $\eta > 1$!

To resolve the first issue, it should be noted that an efficiency >100% in any fuel-to-power conversion system is in violation of the conservation of energy. For further elaboration, let us reconsider Fig. 9.5. As discussed previously, the maximum efficiency monotonically increases with Λ and reaches the 100% limit at a certain value of Λ. The results for the maximum efficiency in Fig. 9.5 are calculated at the regime of fully reversible, i.e., the total entropy generation rate is zero.

It is mathematically correct that a further increase in Λ beyond the 100% efficiency limit (e.g., $\Lambda = 7.2$ in hydrogen-air fuel cell) would yield a maximum power output greater than the total input energy provided we continue to assume that the system still operates at the reversible limit. However, since an efficiency of >100% violates the energy conservation principle, we conclude that the reversible operation would no longer be possible, and the generation of entropy would be unavoidable. For instance, it is impossible for the hydrogen-air fuel cell to operate at fully reversible limit for any value of Λ >7.2 (see Fig. 9.5) so the maximum efficiency may approach 100% while the system produces entropy.

9.4.2 Second issue

Shown in Fig. 9.5 is also the efficiency of a Carnot engine operating between the same low temperature ($T_R = 298.15$ K) and the high temperature (T_P) of the fuel cell. It is obvious that for each mole of hydrogen consumed in the cell and fixed reactants temperature, an increase in the air amount supplied to the cell would lower the temperature of the reaction products, hence the corresponding Carnot efficiency would also decrease. Fig. 9.5 clearly shows

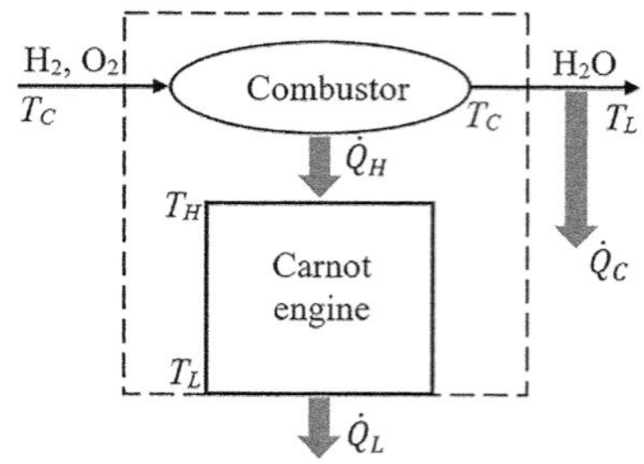

Fig. 9.6 A model of combined Carnot engine and an isothermal combustor.

that under identical operating conditions, the efficiency of a fuel cell may become greater than the Carnot efficiency.

Fig. 9.6 depicts a model of combined Carnot engine and an isothermal combustor, which is often used in some arguments to justify that the upper limit of the fuel cell efficiency is the Carnot efficiency. Hydrogen and oxygen are fed to the combustor at temperature T_C, and water exiting the combustor is discharged to the surrounding. The model of Fig. 9.6 is very much like the power plant model of Bejan [5] with a difference that the combustor in Fig. 9.6 is isothermal whereas that presented by Bejan is adiabatic.

To reversibly transfer heat from the combustor to the engine dictates $T_H = T_C$, and because $T_H > T_L$ we conclude that the combustion product (water) leaves the combustor at T_H. Since the water is discharged to the surrounding, as in most real engines, we need to account for the entropy generation due to the cooling process of the combustion product from T_H to T_L. Thus, the model of Fig. 9.6 is not fully reversible, and the irreversibility sources shown in Fig. 9.6 should be included in the calculation of the total entropy generation rate. Hence,

$$\dot{\Phi} = \frac{\dot{Q}_L}{T_L} + \frac{\dot{Q}_C}{T_L} + \left[\dot{n}_{H_2O}(s_{H_2O})_{T_L} - (\dot{n}_{H_2}s_{H_2} + \dot{n}_{O_2}s_{O_2})_{T_H}\right] \tag{9.25}$$

On the other hand, the thermal efficiency of the engine model of Fig. 9.6 is

$$\eta = \frac{\dot{W}_C}{\dot{Q}_{in}} \tag{9.26}$$

where

$$\dot{W}_C = \dot{Q}_H \eta_C = -\Delta H \left(1 - \frac{T_L}{T_H}\right) \tag{9.27}$$

$$\dot{Q}_{in} = \dot{W}_C + \dot{Q}_L + \dot{Q}_C \tag{9.28}$$

and $-\Delta H$ is the reaction heat evaluated at T_H.

For a fixed T_L, both the total entropy generation rate and the thermal efficiency are a function of T_H only. Fig. 9.7 displays the thermal efficiency, the efficiency of the Carnot engine, and the specific entropy generation (SEG), i.e., $\dot{\Phi}/\dot{n}_{H_2}$ (see Chapter 8) against T_H. It is evident that the thermal efficiency of the model of Fig. 9.6 is less than the corresponding Carnot efficiency over the temperature range shown in Fig. 9.7. The efficiency of the Carnot engine monotonically increases with T_H. On the other hand, the system efficiency attains a maximum at the same optimum T_H, which minimizes the SEG. The inverse relation between the thermal efficiency and SEG is presented in detail in Chapter 8.

Comparing the efficiency of the hydrogen-oxygen fuel cell in Fig. 9.4 with that of the Carnot engine in Fig. 9.7, it can be readily deduced that limiting the fuel cell efficiency to the Carnot efficiency operating between the same low and high temperature is incorrect from thermodynamic perspective. For instance, at a temperature of 1000 K, the efficiency of H_2-air

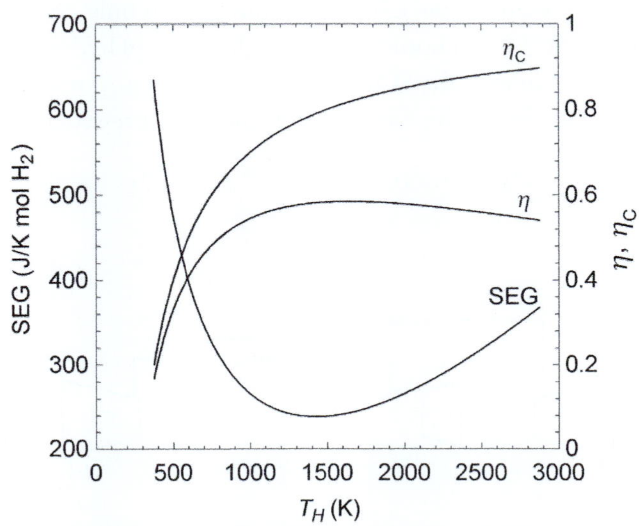

Fig. 9.7 Variation of the thermal efficiency, Carnot efficiency, and SEG with T_H ($T_L = 298$ K).

fuel cell is 0.795 whereas that of the Carnot engine is 0.7. It is important to realize that an ideal fuel cell with an efficiency greater than the Carnot efficiency is not in violation of the second law; a natural law that prohibits complete conversion of *heat* to mechanical work.

9.5 SEG in a hybrid cycle

In this last section, the application of the SEG analysis is demonstrated in a hybrid power generating system. Fig. 9.8 shows schematically a combined gas turbine and solid oxide fuel cell (SOFC) cycle. The fuel is hydrogen and air is used as the oxidizer. Both the air and hydrogen supplied to the system at the ambient temperature and pressure are compressed through individual compressors up to a desired level. In practice, only a fraction of the fuel is consumed within the fuel cell, which is taken into account through a *fuel utilization factor* denoted by U_f. In a hybrid cycle like that shown in Fig. 9.8, the unused portion of the fuel is burned in a downstream combustor. Like in conventional power plants, the amount of air supplied to the cathode is well above the minimum stochiometric value. The additional air supplied to the cathode is accounted for by an *air utilization factor* represented by U_a.

9.5.1 SOFC model

Since the thermodynamic model of the components of gas turbine cycles has previously been presented (see Chapter 8), this section only describes a simplified model for SOFC. Denoting the molar flowrates of hydrogen, oxygen, and nitrogen at the inlet of the SOFC by $\dot{n}_{H_2}$, $\dot{n}_{O_2}$, and $\dot{n}_{N_2}$, respectively, the total reactants enthalpy at the fuel cell inlet can be expressed as

$$H_R = \dot{n}_{H_2} h_{H_2} + \left(\dot{n}_{O_2} h_{O_2} + \dot{n}_{N_2} h_{N_2}\right)_2 \tag{9.29}$$

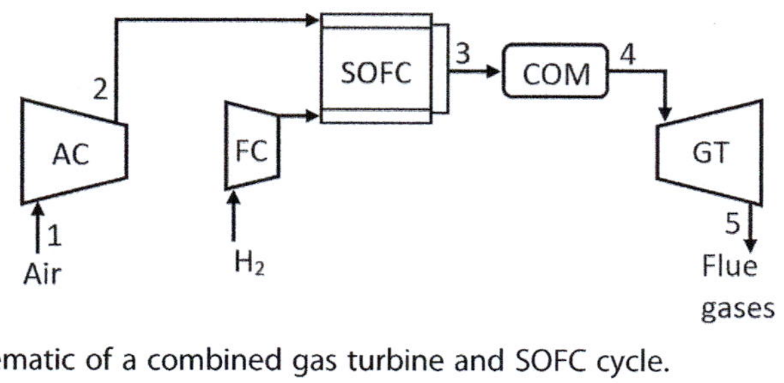

Fig. 9.8 Schematic of a combined gas turbine and SOFC cycle.

Likewise, the total enthalpy of the products at the exit of the SOFC (state 3) is

$$H_P = \left[\dot{n}_{H_2}(1-U_f)h_{H_2} + \dot{n}_{O_2}h_{O_2} + \dot{n}_{N_2}h_{N_2} + \dot{n}_{H_2O}h_{H_2O}\right]_3 \tag{9.30}$$

Applying an energy balance to the SOFC operating under adiabatic and steady-state conditions gives

$$H_R = \dot{W}_{FC,dc} + H_P \tag{9.31}$$

The DC (direct current) electric power production of the SOFC is related to the operating voltage and current of the cell.

$$\dot{W}_{FC,dc} = V \cdot I \cdot N_c \tag{9.32}$$

where N_c is the number of the cells.

The molar flowrates of the hydrogen and oxygen fed to the anode are obtained using Eqs. (9.33) and (9.34), respectively.

$$\dot{n}_{H_2} = \frac{IN_c}{2F}\left(\frac{1}{U_f}\right) \tag{9.33}$$

$$\dot{n}_{O_2} = \frac{IN_c}{4F}\left(\frac{1}{U_a}\right) \tag{9.34}$$

Given the O_2/N_2 ratio of the air, one may determine the molar flowrate of the nitrogen as

$$\dot{n}_{N_2} = \frac{0.79}{0.21}\dot{n}_{O_2} \tag{9.35}$$

For each mole of hydrogen participated in the cell reaction, one mole water is formed. So, the molar flowrate of the water at the SOFC exit is obtained by

$$(\dot{n}_{H_2O})_3 = U_f \dot{n}_{H_2} \tag{9.36}$$

On the other hand, half mole of oxygen participates in the cell reaction per each mole of hydrogen. So, the oxygen flowrate at the SOFC exit is

$$(\dot{n}_{O_2})_3 = \frac{IN_c}{4F}\left(\frac{1}{U_a} - 1\right) \tag{9.37}$$

Because nitrogen does not participate in the reaction, its flowrate at the inlet and outlet of the fuel cell is the same.

Next, we evaluate the entropy generation rate of the SOFC. The total entropy flow at the fuel cell inlet is

$$S_2 = \dot{n}_{H_2} s_{H_2} + \left(\dot{n}_{O_2} s_{O_2} + \dot{n}_{N_2} s_{N_2}\right)_2 \tag{9.38}$$

Likewise, the total enthalpy flow of the reaction products at state 3 is

$$S_3 = \left[\dot{n}_{H_2}\left(1 - U_f\right) s_{H_2} + \dot{n}_{O_2} s_{O_2} + \dot{n}_{N_2} s_{N_2} + \dot{n}_{H_2O} s_{H_2O}\right]_3 \tag{9.39}$$

The specific entropy of the individual species in Eq. (9.39) should be calculated at temperature T_3 and their partial pressure, p_i, at state 3, where

$$p_i = p_3 \left(\frac{\dot{n}_i}{\dot{n}_{tot}}\right)_3 \quad i: H_2, O_2, N_2, H_2O \tag{9.40}$$

The entropy generation rate of the SOFC is determined by

$$\dot{\Phi}_{SOFC} = S_3 - S_2 \tag{9.41}$$

The total molar flowrate of the SOFC products flowing to the combustor can be determined using Eq. (9.33) through Eq. (9.37). Hence,

$$\left(\dot{n}_{tot}\right)_3 = \frac{IN_c}{2F}\left(\frac{1}{U_f} + \frac{4.76}{2U_a} - \frac{1}{2}\right) \tag{9.42}$$

The unburned portion of the hydrogen is assumed to completely oxidize within the combustor.

9.5.2 Illustrative example

A numerical example is now presented to examine the performance of the hybrid system of Fig. 9.8 using the following operating parameters. The SOFC stack consists of two cells each with a surface area of $1000\,\text{cm}^2$. The operating voltage, current density, and temperature are 0.68 V, $300\,\text{mA/cm}^2$, and 1173 K, respectively. The fuel utilization factor is assumed to be 0.8. The isentropic efficiencies of the gas turbine and compressors are 0.90 and 0.85, respectively. Both the hydrogen and air are supplied to the cycle at 298.15 K and 1 bar.

The net power generation and the thermal efficiency of the hybrid cycle are obtained as follows.

$$\dot{W}_{net} = \dot{W}_{FC,ac} + \dot{W}_t - \dot{W}_c - \dot{W}_{fc} \tag{9.43}$$

$$\eta = \frac{\dot{W}_{net}}{\dot{n}_{H_2} LHV} \tag{9.44}$$

where

$$\dot{W}_{FC,ac} = \eta_{inv} \dot{W}_{FC,dc} \tag{9.45}$$

and η_{inv} denotes the DC-AC inverter efficiency assumed to be 0.95 [6] in the present calculations.

Fig. 9.9 depicts the thermal efficiency and the specific entropy generation of the hybrid cycle varying with the compressor pressure ratio. The total entropy generation rate of the cycle is the sum of the entropy generation rates within the components of the cycle and due to the cooling of the flue gases discharged to the atmosphere. Hence,

$$\dot{\Phi} = \dot{\Phi}_c + \dot{\Phi}_{fc} + \dot{\Phi}_{SOFC} + \dot{\Phi}_{com} + \dot{\Phi}_t + \dot{\Phi}_L \tag{9.46}$$

The calculation of the entropy generation due to the heat rejection to the surrounding, $\dot{\Phi}_L$, is explained in Chapter 8; see Sections 8.4.2, 8.5.1, and 8.6.1. The specific entropy generation of the cycle is simply obtained using Eq. (9.47).

$$\text{SEG} = \frac{\dot{\Phi}}{\dot{n}_{H_2}} \tag{9.47}$$

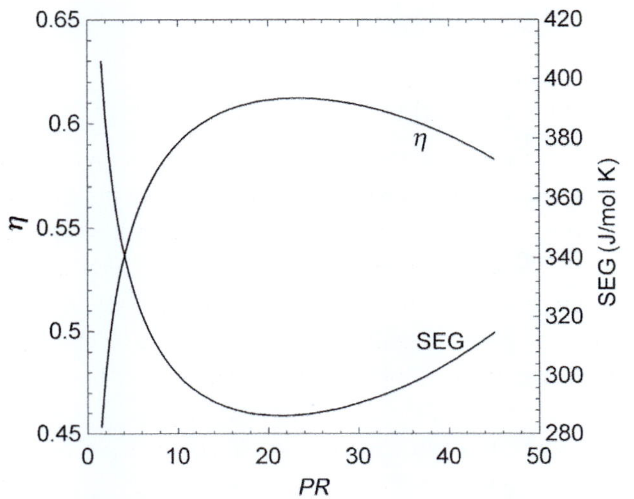

Fig. 9.9 Variation of the thermal efficiency and SEG of the integrated gas turbine and SOFC cycle with the compressor pressure ratio.

Like the conventional power plants discussed in Chapter 8, it can be readily deduced from Fig. 9.9 that the maximum thermal efficiency design is identical to that of the minimum SEG for the hybrid cycle. The optimum pressure ratio in Fig. 9.9 is ~21.7 at which the cycle efficiency is 0.612 (i.e., 61.2%). The distribution of the SEG within the cycle at the maximum efficiency operation is shown in Fig. 9.10. The exhaust gases, with 46% share, are the largest source of the efficiency losses, followed by the SOFC with 30% contribution to the inefficiencies.

The temperature of the exhaust gases leaving the turbine is relatively high, 735.3 K, which explains why the exhaust gases are responsible for a significant portion of the efficiency losses. A modified design may then include a recuperator to partially recover the thermal energy of the hot exhaust gases by preheating the air upstream of the SOFC. The modified cycle would indeed be an integrated *regenerative* gas turbine and SOFC cycle. Assuming an effectiveness of 0.85 for the recuperator, the maximum thermal efficiency raises to 0.658. This is 4.6 percentage points higher than the maximum efficiency of the original hybrid cycle of Fig. 9.8. The key features of the original and the modified hybrid cycles are compared in Table 9.3.

The optimum pressure ratio of the regenerative hybrid cycle, i.e., 5.6, is almost a quarter of that without a recuperator. The exhaust gases in the modified cycle are over 120 K cooler than those in the hybrid cycle of Fig. 9.8.

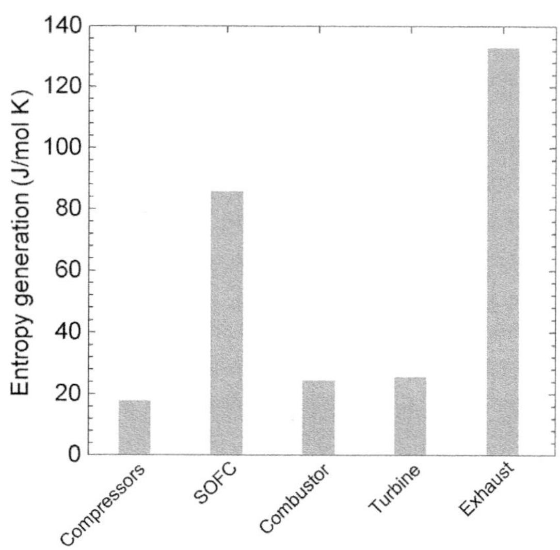

Fig. 9.10 Distribution of the SEG within the cycle at maximum efficiency.

Table 9.3 Performance comparison of the integrated gas turbine and SOFC with and without recuperator at maximum thermal efficiency.

Parameter	With recuperator	Without recuperator
Optimum pressure ratio	5.6	21.7
TIT (K)	1347	1384
Exhaust temperature (K)	614.6	735.3
Net power production (kW)	618.3	575.2
Thermal efficiency	0.658	0.612
SEG (J/mol K)	251.1	286.2
Air flowrate (mol/s)	29.41	23.56

The heat integration allows to augment the thermal efficiency from 0.612 to 0.658 at the optimum operation. This gain in the thermal efficiency corresponds to 35.1 J/mol H_2·K reduction in the SEG.

References

[1] J. Larminie, A. Dicks, Fuel Cell Systems Explained, second ed., Wiley, West Sussex.
[2] Eg&G Technical Services Inc., Fuel Cell Handbook, seventh ed., US Department of Energy.
[3] P.E. Dodds, I. Staffell, A.D. Hawkes, F. Li, P. Grünewald, W. McDowall, P. Ekins, Hydrogen and fuel cell technologies for heating: a review, Int. J. Hydrogen Energy 40 (2015) 2065–2083.
[4] A.E. Lutz, R.S. Larson, J.O. Keller, Thermodynamic comparison of fuel cells to the Carnot cycle, Int. J. Hydrogen Energy 27 (2002) 1103–1111.
[5] A. Bejan, Models of power plants that generate minimum entropy while operating at maximum power, Am. J. Phys. 64 (1996) 1054–1059.
[6] S. Wongchanapai, H. Iwai, M. Saito, H. Yoshida, Performance evaluation of a direct-biogas solid oxide fuel cell-micro gas turbine (SOFC-MGT) hybrid combined heat and power (CHP) system, J. Power Sources 223 (2013) 9–17.

CHAPTER TEN

Entropy and chemical equilibrium

10.1 Introduction

Many practical applications involve complex processes with numerous chemical reactions, e.g., combustion of fuel, gasification of biomass, steam reforming of hydrocarbons. The byproducts of the reactions often consist of several substances. The product of biomass gasification, as an example, is a mixture of mainly H_2, CO, CO_2, CH_4, and H_2O with trace of other heavy hydrocarbons such as C_2H_4, C_2H_6, C_6H_6 [1]. To correctly and accurately predict the gasification process, one needs to account for various transport mechanisms coupled with the kinetics of the reactions involved. Development of a mathematical model that includes these effects and the numerical solution of such models is a challenging task, but the outcome is rewarding.

A solution of a coupled transport-kinetic model applied to a reactive flow allows prediction of species concentrations, temperature, and velocity along the flow path. For instance, the predicted composition of the gasification products, also called producer gas, using such models is shown to be highly accurate [2,3]. It is important to realize that the transport-kinetic models rest on the laws of nature—the primary reason for their high accuracy. The transport equations are indeed analytical expressions of the conservation laws. The kinetic models that are employed to predict the rates of formation or consumption of chemical compounds are also based on experimental observations.

Now the question is whether one may determine the composition of a chemical reaction at the state of chemical equilibrium by mere laws of thermodynamics. This problem was investigated in the late 19th century by Gibbs, and later by Nernst who examined the theory of Gibbs through a limited number of experiments. The criterion proposed by Gibbs postulates that at chemical equilibrium the Gibbs energy of a system maintained at fixed temperature and pressure is a minimum.

$$dG = 0 \tag{10.1}$$

Entropy Analysis in Thermal Engineering Systems
https://doi.org/10.1016/B978-0-12-819168-2.00010-6

where G denotes the Gibbs function defined as $G = H - TS$, H is the enthalpy of the system, T the temperature, and S the entropy.

A second class of models widely used for prediction of the composition of a chemical reaction employs the conservation of energy, conservation of elements (e.g., carbon, hydrogen, oxygen), and the Gibbs criterion, but they disregard the physiochemical processes. The main task of this chapter is to evaluate the validity of Eq. (10.1).

10.2 Definition of equilibrium

Equilibrium refers to a static state, which, if disturbed, will lead to a flow or motion. Fig. 10.1 displays three examples of equilibrium. Thermal equilibrium, as shown in Fig. 10.1A, refers to a uniform temperature across a system or between a system and its surrounding. It can be disturbed by creating a temperature gradient between two different locations of the system; this will lead to a flow of heat. Fig. 10.1B illustrates a static equilibrium. The summation of all external forces acting on a body should be zero at static

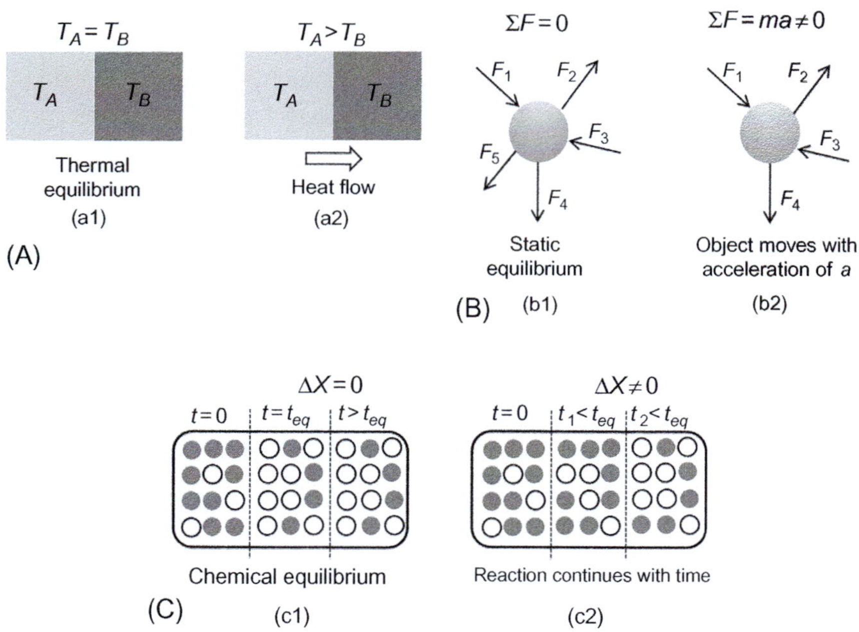

Fig. 10.1 An illustrative description of (A) thermal equilibrium, (B) static equilibrium, and (C) chemical equilibrium.

equilibrium. An imbalance between the external forces would initiate motion of the body.

Likewise, chemical equilibrium is the state of a reactive system at which the quantity (e.g., mole, concentration) of the individual components (reactants and products) does not vary with time. Fig. 10.1c1 shows a reactive system that initially ($t = 0$) contains reactants (filled circles) and products (empty circles). The system approaches equilibrium at $t = t_{eq}$ beyond which the composition of the species will no longer be dependent on time, so all reactants and products will have uniform concentrations. As depicted in Fig. 10.1c2, between $t=0$ and $t=t_{eq}$, the quantity of individual species may continuously change in time.

For a reactive system comprising k species, the condition of chemical equilibrium may analytically be expressed as

$$\Delta X_j(t) = \varepsilon \quad \left(t > t_{eq}\right); \; j = 1, 2, \ldots, k \tag{10.2}$$

where X_j denotes the conversion of species j, and ε is a negligible real number.

Eq. (10.2) states that after a chemical equilibrium has been established in a reactive system, the change in the quantity of any species like j is negligible at any instant beyond the equilibrium time, t_{eq}. One may also express the condition of chemical equilibrium described in Eq. (10.2) in terms of concentration C, mole fraction y, or mass fraction x, i.e.,

$$\Delta C_j(t) = \varepsilon_1 \quad \left(t > t_{eq}\right) \tag{10.2a}$$

$$\Delta y_j(t) = \varepsilon_2 \quad \left(t > t_{eq}\right) \tag{10.2b}$$

$$\Delta x_j(t) = \varepsilon_3 \quad \left(t > t_{eq}\right) \tag{10.2c}$$

Whether the state of chemical equilibrium characterized by Eq. (10.2) may also be determined by minimization of Gibbs function, Eq. (10.1), is a question that we aim to answer in the following sections.

10.3 Experimental examination of theory

Equilibrium-based modeling approaches have been used widely, mainly due to their simplicity compared to the coupled transport-kinetic models, for prediction of the composition of reactive mixtures. In general, an equilibrium model consists of three main parts: (i) conservation of elements, e.g. carbon, hydrogen, oxygen, (ii) conservation of energy assuming a uniform temperature, and (iii) Gibbs criterion of equilibrium, i.e., Eq. (10.1). The energy and mass conservations are two fundamental laws of nature whose validity has been supported by experiments. The Gibbs criterion, however, rests heavily on theoretical arguments.

The predictability of the equilibrium-based models is examined by reviewing the scholarly reports where the composition of the reaction products is calculated by minimization of Gibbs function. The survey has revealed significant discrepancies, both quantitatively and qualitatively, between the equilibrium calculations and experimental measurements in several studies. Table 10.1 provides a comparison between the equilibrium model prediction and measured compositions of the products of biomass gasification. These data are collected from Refs. [4–12].

An average discrepancy between the measured and predicted data of Table 10.1 is calculated for the gaseous species and depicted in Fig. 10.2. The traditional equilibrium-based estimated nitrogen, carbon dioxide, and carbon monoxide yields are within ±10% of the experimental values. However, the hydrogen yield is 57% overpredicted, whereas that of methane is roughly 90% underpredicted. The failure of the equilibrium-based models to correctly predict the compositions of biomass gasification can also be found in Refs. [13–15].

The discrepancy between the experiments and predictions of the equilibrium models is commonly reasoned to be due to the reaction products

Table 10.1 Producer gas compositions obtained from biomass gasification.

Species	Experiment	Model	Notes	Source
H_2	12.5	18.03	MC: 14%	Jarungthammachote and Dutta [4]
CH_4	1.2	0.11	Temperature: 1273 K	
CO	18.9	18.51	Feedstock: Rubber wood	
CO_2	8.5	11.43	AFR: 2.29	
N_2	59.1	51.92	Gasifier: Downdraft	
H_2	8.43	14.99	Temperature: 1148.7 K	Jarungthammachote and Dutta [5]
CH_4	2.52	0	Feedstock: Coconut shell	
CO	11.61	20.68	ER: 0.35	
CO_2	14.95	10.42	Gasifier: Spout fluid bed	
N_2	61.55	53.9		
H_2	14.1	21.0	Feedstock: Sawdust	Ruggiero and Manfrida [6]
CH_4	3.5	1.3	Gasifier: Circulating fluidized bed	
CO	18.7	13.2		
CO_2	14.7	15.8		

Table 10.1 Producer gas compositions obtained from biomass gasification.—cont'd

Species	Experiment	Model	Notes	Source
N_2	47.7	45.1		
H_2	9.5	19.4	Feedstock: Sawdust	
CH_4	4.5	0.9	Gasifier: Bubbling fluidized bed	
CO	18.0	22.0		
CO_2	13.5	11.4		
N_2	45.0	45.1		
H_2	15.23	21.06	MC: 20%	Zainal et al. [7]
CH_4	1.58	0.64	Temperature: 1073 K	
CO	23.04	19.61	Feedstock: Unspecified wood	
CO_2	16.42	12.01		
N_2	42.31	46.68		
O_2	1.42	0		
H_2	14.0	20.39	MC: 10%	Renkel and Lümmen [8]
CH_4	2.31	0.02	Temperature: 1073 K	
CO	20.14	21.78	Feedstock: Sawdust	
CO_2	12.06	10.38	Gasifier: Downdraft	
N_2	50.79	47.44		
H_2	43.2	58	MC: 7.5%	Mahishi and Goswami [9]
CH_4	7.4	0	Temperature: 1123 K (ER: 0.09)	
CO	18.5	19.8		
CO_2	28.5	22.2	Feedstock: Sawdust Gasifier: Fluidized bed	
H_2	27.6	48.5	MC: 7.5%	
CH_4	7.7	0	Temperature: 1123 K (ER: 0.37)	
CO	15.5	16.5		
CO_2	46.9	35	Feedstock: Sawdust Gasifier: Fluidized bed	
H_2	12.5	17.9	MC: 14%	Fournell et al. [10]
CH_4	1.2	0	Temperature: 1273 K	
CO	18.9	19.4	Feedstock: Rubber wood	
CO_2	8.5	10.9	AFR: 2.29	
N_2	59.1	51.8	Gasifier: Downdraft	
H_2	14.0	20.06	MC: 10%	Altafini et al. [11]
CH_4	2.31	0	Temperature: 1073 K	
CO	20.14	19.7	Feedstock: Pinus Elliotis	
CO_2	12.06	10.15	Gasifier: Downdraft	
N_2	50.79	50.1		
H_2	5.6	9.9	MC: 22%	Baratieri et al. [12]
CH_4	1.4	0.002	Temperature: 1013 K	
CO	6.9	7.6	Feedstock: Cypress	
CO_2	18.1	17	AFR: 0.29	
N_2	68	65.2	Gasifier: Circulating fluidized bed	

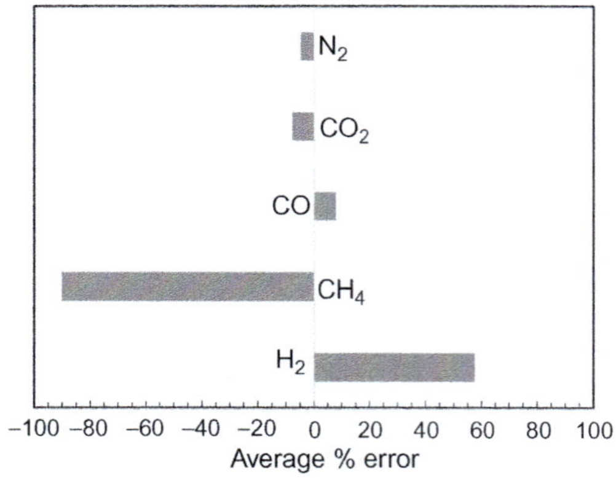

Fig. 10.2 Average percentage error between the predicted and measured compositions of the producer gas given in Table 10.1.

being far from a chemical equilibrium. The quantitative disagreement between the predicted and measured values discussed above is not the only concern as in some studies [16,17] a qualitative mismatch has also been observed. In a study of water-gas shift reactor [16], the CO conversion was accurately predicted with a kinetic-based detailed model. However, the prediction of the equilibrium model was unsatisfactory as the trend of CO conversion vs temperature was opposite to that of the experiment. It deems therefore necessary to continue investigating the basis and validity of the traditional criterion of equilibrium from a fundamental perspective, which is the subject of the next section.

10.4 Thermodynamics of chemical reaction

Consider a system of k different substances maintained at a uniform temperature T and pressure p with a known initial composition.

$$n^i = n_1^i + n_2^i + \ldots + n_k^i = \sum_{j=1}^{k} n_j^i \tag{10.3}$$

where n is the number of moles and superscript i designates the initial state.

The total enthalpy and entropy of the system at the initial state are determined by Eqs. (10.4) and (10.5), respectively.

$$H^i = \sum_{j=1}^{k} \left[n_j h_j(T)\right]^i \tag{10.4}$$

$$S^i = \sum_{j=1}^{k} \left[n_j s_j(T, p)\right]^i \tag{10.5}$$

where h is the specific molar enthalpy, and s denotes the specific molar entropy.

If the system undergoes a chemical reaction, its composition at the final state will be different than the initial composition. Hence,

$$n^f = n_1^f + n_2^f + \ldots + n_k^f = \sum_{j=1}^{k} n_j^f \tag{10.6}$$

10.4.1 Exothermic reaction

In this case, the reaction leads to a release of thermal energy. If the process takes place adiabatically, the temperature of the system will increase. Suppose that the heat Q is transferred from the system to the surrounding that is at temperature T_s. The amount of Q sufficient to reduce the temperature of the system to the initial temperature is obtained by applying the first law. Hence,

$$\sum_{j=1}^{k} \left[n_j h_j(T)\right]^i = Q + \sum_{j=1}^{k} \left[n_j h_j(T)\right]^f \tag{10.7}$$

The total entropy generation is determined as

$$\Phi = \sum_{j=1}^{k} \left[n_j s_j(T, p)\right]^f - \sum_{j=1}^{k} \left[n_j s_j(T, p)\right]^i + \frac{Q}{T_s} \tag{10.8}$$

The first two bracketed terms on the right-hand side of Eq. (10.8) account for the change in the entropy of the system, whereas the last term is the net increase in the entropy of the surrounding. A combination of Eqs. (10.7) and (10.8) yields

$$\Phi = -\frac{1}{T_s}\left\{\sum_{j=1}^{k} n_j^f \left[h_j(T) - T_s s_j(T, p)\right]^f - \sum_{j=1}^{k} n_j^i \left[h_j(T) - T_s s_j(T, p)\right]^i\right\} \tag{10.9}$$

Introducing a new function $g_m = h(T) - T_s s(T, p)$, Eq. (10.9) may be represented as

$$\Phi = -\frac{1}{T_s}\left(\sum_{j=1}^{k} G_{m,j}^{f} - \sum_{j=1}^{k} G_{m,j}^{i}\right) = -\frac{\Delta G_m}{T_s} \tag{10.10}$$

where $G_{m,j} = n_j g_{m,j}$, and ΔG_m denotes the change in G_m due to the reaction.

10.4.2 Endothermic reaction

An endothermic reaction requires heat to be supplied from an external source. The amount of heat Q that should be transferred to the system from a heat source that is at temperature T_s to maintain an isothermal reaction can be determined using the first law.

$$\sum_{j=1}^{k} \left[n_j h_j(T)\right]^i + Q = \sum_{j=1}^{k} \left[n_j h_j(T)\right]^f \tag{10.11}$$

The total entropy generation associated with the isothermal reaction is obtained by

$$\Phi = \sum_{j=1}^{k} \left[n_j s_j(T,p)\right]^f - \sum_{j=1}^{k} \left[n_j s_j(T,p)\right]^i - \frac{Q}{T_s} \tag{10.12}$$

The negative sign of Q/T_s denotes the reduction in the entropy of the heat source. Combining Eqs. (10.11) and (10.12) to eliminate Q would again lead to Eq. (10.9). So, Eqs. (10.9) and (10.10) are valid for both endothermic and exothermic reactions. It must be remembered that the formulation rests on the key assumption of constant T_s.

10.4.3 Gibbs function

An essential element in the arguments of Gibbs is the condition of reversibility without altering temperature. For a system undergoing a non-isothermal process, he postulated that "it is not necessary ... that the temperature of the system should remain constant during the reversible process, ... provided that the only source of heat or cold used has the same temperature as the system in its initial or final state. Any external bodies may be used in the process in any way not affecting the condition of reversibility." He then elucidated "uniformity of temperature and pressure are always necessary for equilibrium, and the remaining conditions, when these are satisfied, may be conveniently expressed by means of the function ζ" [18], where ζ is indeed the Gibbs function G.

Gibbs's theory is concerned with reversible processes with the surrounding temperature assumed to be the same as the system temperature. Setting $T_s = T$ in our formulation yields $G_m = G$, and Eq. (10.10) is accordingly expressed in terms of the Gibbs function.

$$\Phi_G = -\frac{1}{T}\left(\sum_{j=1}^{k} G_j^f - \sum_{j=1}^{k} G_j^i\right) = -\frac{\Delta G}{T} \tag{10.13}$$

where Φ_G is the entropy generation function that is evaluated for the special condition of $G_m = G$, and ΔG denotes the net difference in the Gibbs function between the initial and final states.

Because real processes are irreversible, we have $\Phi > 0$ so the right-hand side of Eq. (10.10) is always positive. Thus, the change in function G_m will always be negative, i.e., $\Delta G_m < 0$. Hence,

$$G_m^f < G_m^i \tag{10.14}$$

In the special case of $T_s = T$, one finds $G^f < G^i$. Inequality (10.14) is obtained by invoking the second law. It is valid for all thermodynamic processes in the absence of work, including chemical reactions, which exchange heat with the surroundings assumed to be at a constant temperature. It states that for a system undergoing an actual process that is initially at an arbitrary but known state, the function G_m at the final state will always be less than that in the initial state regardless of whether the transformation from the initial state to the final state is infinitesimal, the system is initially at equilibrium or nonequilibrium, or the surrounding temperature, T_s, is equal to the system temperature. Any other conclusion drawn from inequality (10.14) relating the state of chemical equilibrium to the minimum of G_m (or G in the special case of $T_s = T$) would simply be unjustified.

From the discussion thus far, one may deduce that an application of Gibbs criterion, Eq. (10.1), may lead to inaccurate predictions of the composition of a reactive system because there exists no strong experimental nor analytical foundation for such criterion. Neither the second law nor the combined first and second laws suggest that the chemical equilibrium is designated by setting the differential of the Gibbs function or G_m to zero. In a thermodynamic equilibrium, there is no change in the quantity of any species of the system, enthalpy, entropy, etc. The source of mistake is that "the change in enthalpy or entropy is zero" is often interpreted as "the differential of enthalpy or entropy is zero." One is then led to look for an *extremum* that is erroneously believed to represent the state of chemical equilibrium.

10.5 Reaction advancement

For a reactive system that is at constant temperature T and pressure p, a change in the state of the system could occur due to a change in its composition. The entropy generation Φ and function G_m may then be described in terms of the extent of reaction ξ (also called reaction advancement) [19]. Once a chemical equilibrium is established, we have $\xi(t_{eq})=\xi_{eq}$, where ξ_{eq} denotes the extent of reaction at equilibrium. For $t>t_{eq}$, both $\Phi(\xi)$ and $\Delta G_m(\xi)$ will remain unaltered. This, however, should not be confused or interpreted as $\Phi(\xi)$ and $G^f_m(\xi)$ attaint their extremum at the state of equilibrium.

In a chemically reactive system comprising k different species with a known initial composition, the number of moles of all species participating in the chemical reaction will change with the reaction advancement and may be determined using Eq. (10.15).

$$n^f_j = n^i_j + a_j\xi \tag{10.15}$$

where a_j is the stochiometric coefficient of species j that takes part in the reaction.

Substituting Eq. (10.15) into Eq. (10.6), one obtains

$$n^f = \sum_{j=1}^{k} n^f_j = \sum_{j=1}^{k}\left(n^i_j + a_j\xi\right) \tag{10.16}$$

Using Eq. (10.3), one may simplify Eq. (10.16) to derive a relation for the total moles of the system at the final state.

$$n^f = n^i + a\xi \tag{10.17}$$

where

$$a = \sum_{j=1}^{k} a_j \tag{10.18}$$

Because the initial state of the system is fixed, a change in function G_m solely depends on its value at the final state, which can be described in terms of the reaction advancement as

$$G_m^f(\xi) = \sum_{j=1}^{k} G_{m,j}^f(\xi) = \sum_{j=1}^{k} \left(n_j^i + a_j\xi\right)\left[h_j(T) - T_s s_j(T,p)\right]^f \tag{10.19}$$

For a reactive system with a fixed initial state, Eq. (10.19) states that G_m^f is only a function of ξ. Also, since G_m^i is constant, the entropy generation $\mathbf{\Phi}$ is dependent on G_m^f—this can be deduced from Eq. (10.10). We therefore conclude that $\mathbf{\Phi}$ is also a function of ξ only.

10.5.1 Methane steam reforming

Let us now consider a simple example where the reactants are methane and steam and the products are hydrogen and carbon monoxide.

$$CH_4 + H_2O \rightarrow 3H_2 + CO \tag{10.20}$$

The reaction takes place at an elevated temperature T and 1 bar. The stochiometric coefficients of reaction (10.20) are: $a_{CH_4} = -1$, $a_{H_2O} = -1$, $a_{H_2} = 3$, and $a_{CO} = 1$. From Eq. (10.18), we then find $a = 2$. Fig. 10.3 depicts the variation of the change in the thermodynamic function G_m^f and entropy generation with ξ for the methane steam reforming reaction assuming 3 kmol methane and 5 kmol steam at the initial state. Since the reforming reaction is endothermic, T_s represents the temperature of the heat source. The graphs in Fig. 10.3 are obtained for a reaction temperature of 800 K

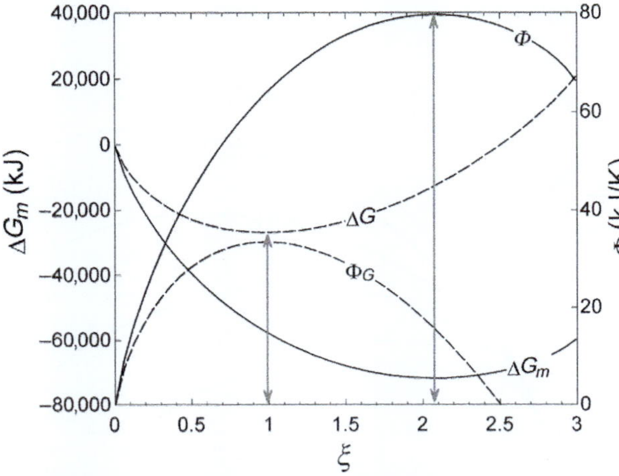

Fig. 10.3 Variation of the entropy generation and ΔG_m with the extent of reaction for methane steam reforming at 800 K and 1 bar, $T_s = 900$ K. The graphs of Φ_G and ΔG correspond to the special case of $T_s = T$.

and assuming $T_s = 900\,K$. The dashed lines represent the special case of Gibbs, where $T_s = T$ so that $G_m = G$ and $\Phi = \Phi_G$; see Section 10.4.3.

The maximum entropy production and the minimum change in G_m coincide at $\xi = 2.08$. As the initial state is fixed, G_m^i is independent of ξ so only G_m^f is a function of ξ. The minimum ΔG_m corresponds to the minimum value of G_m^f. On the other hand, the maximum entropy production and the minimum ΔG (dashed lines in Fig. 10.3) coincide at $\xi \approx 1$, which is less than the optimum ξ that maximizes Φ and minimizes G_m^f. From thermodynamic perspective, in the special case of Gibbs, the reaction may proceed up to a maximum value of $\xi_{max} \approx 2.5$ at which $\Phi_G(\xi_{max}) = \Delta G(\xi_{max}) = 0$. The reaction cannot take place for $\xi > 2.5$ which would otherwise violate the second law. The reaction could only proceed for $0 < \xi \leq 2.5$. In reality, $T_s \neq T$ and as seen in Fig. 10.3, Φ is always positive for the entire range of ξ so the reaction may occur at any ξ without violating the laws of thermodynamics.

Interesting to note is that the magnitude of the optimum ξ depends on the surrounding temperature. As shown in Fig. 10.4, at a higher heat source temperature, the location of the maximum entropy production shifts toward a higher reaction advancement. The dashed line denoted with $T_s = 800\ K$ corresponds to the Gibbs function. It can be deduced from Fig. 10.4 that a higher temperature difference between the reacting system and the heat source favors completion of the reaction.

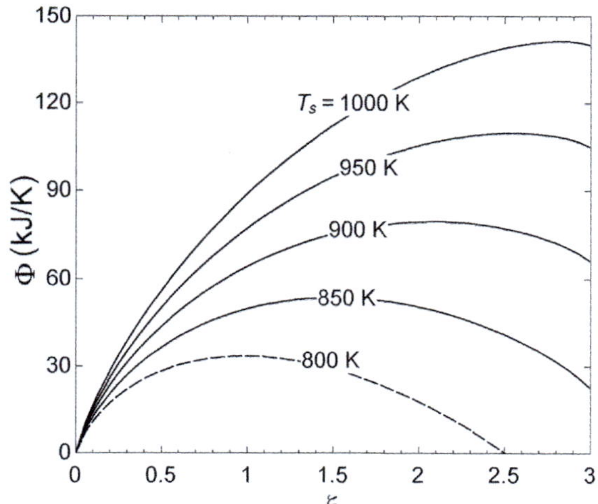

Fig. 10.4 Variation of the entropy generation with the extent of reaction at different heat source temperatures for methane steam reforming at 800 K and 1 bar.

10.5.2 Kinetic model

We now examine the chemical equilibrium of methane steam reforming using the kinetics of reaction. The rate of reaction at which methane is consumed is determined by [20]

$$r = \kappa C_{CH_4} C_{H_2O} \tag{10.21}$$

where C is the concentration and κ denotes the reaction constant that is described in an Arrhenius equation form.

$$\kappa = 3 \times 10^8 \exp\left(-\frac{15{,}000}{T}\right) \left(\text{m}^3/\text{kmol s}\right) \tag{10.22}$$

Fig. 10.5 shows the variation of the mole fractions of the chemical species participating in methane steam reforming with time. In accordance with Eq. (10.2), we may define a criterion for chemical equilibrium as $\Delta n_{CH_4}/n_{CH_4}{}^{i} = 0.01$, which corresponds to 99% conversion of methane. As depicted in Fig. 10.5, the chemical equilibrium is established at 790 s at which 99% of the initial amount of methane is already consumed. On the other hand, the maximum entropy generation occurs at 140 s assuming a heat source temperature of 900 K. It should be noted that the choice of 99% conversion is entirely arbitrary. We could define the equilibrium instant, for example,

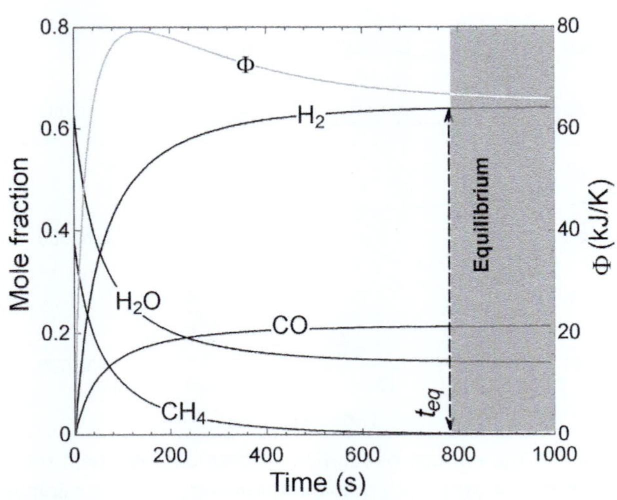

Fig. 10.5 Evolution of the reactants and products of methane steam reforming and the entropy generation at 800 K and 1 bar. The chemical equilibrium is established at $t_{eq} = 790$ s. The maximum entropy generation takes place at 140 s assuming $T_s = 900$ K.

at 98%, 99.99%, etc. conversion. The main idea here is to highlight the physical meaning of the chemical equilibrium defined previously; see Fig. 10.1C.

Fig. 10.5 reveals that calculations performed by maximization of entropy generation (or minimization of ΔG_m) could lead to incorrect prediction of the equilibrium concentrations. However, as evident in Fig. 10.4, the location of maximum entropy production (and thus minimum ΔG_m) shifts toward the completion of the reaction as the temperature of the heat source increases. As depicted in Fig. 10.6, by increasing the heat source temperature from 900 to 1000 K and 1100 K, the time of maximum entropy generation increases from 140 s to 425 s and 880 s, respectively. In other words, only for sufficiently large differences between the heat source temperature and the reaction temperature may the equilibrium composition predicted by maximization of entropy generation be in the vicinity of that obtained by a kinetic model.

Shown in Fig. 10.6 is also the evolution of ΔG whose minimum take place at around 40 s. A comparison of the equilibrium composition (mol%) of the methane steam reforming predicted by the kinetic model, minimization of ΔG_m, and minimization of ΔG is provided in Table 10.2. The results in Table 10.2 further clarify why an equilibrium model based on Gibbs energy minimization may yield inaccurate

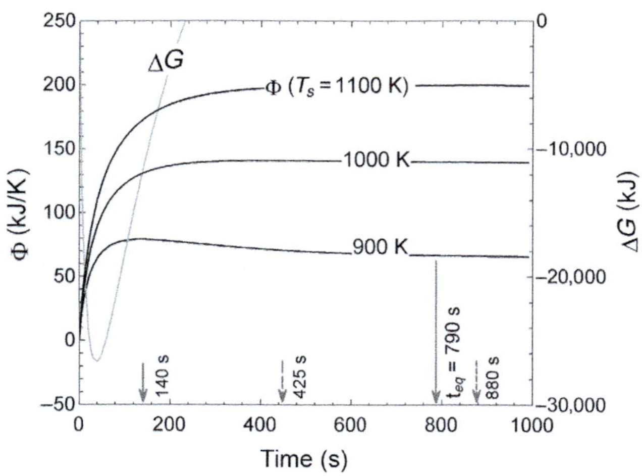

Fig. 10.6 Evolution of the entropy generation at three different heat source temperatures, and ΔG for methane steam reforming at 800 K and 1 bar. The minimum ΔG takes place at around 40 s. For this specific example, the equilibrium compositions predicted by maximization of the entropy generation and the kinetic model are close for $T_s - T \sim 300$ K.

Table 10.2 Comparison of the equilibrium composition (mol%) of methane steam reforming predicted by a kinetic model, minimization of ΔG, and minimization of ΔG_m.

Species	Kinetic	Minimum ΔG	Minimum ΔG_m
H_2O	0.1456	0.3953	0.2377
CH_4	0.0021	0.1964	0.0738
CO	0.2131	0.1021	0.1721
H_2	0.6392	0.3062	0.5164

predictions. Moreover, the prediction of equilibrium by minimization of ΔG_m yields closer results than the Gibbs criterion to the prediction of the kinetic model. Although there is no foundation for relating the state of chemical equilibrium to the minimum of either Gibbs function or G_m, the results presented in Fig. 10.6 and Table 10.2 suggest that calculations preformed using the function G_m are expected to be more accurate than employing the Gibbs function.

A key observation in Figs. 10.4 and 10.5 is that the condition of chemical equilibrium does not necessarily correspond to an extremum of Φ or ΔG_m. Rather, after an establishment of the chemical equilibrium, the values of entropy generation and ΔG_m remain *constant* but not necessarily *optimal*. The condition of chemical equilibrium given by Eq. (10.2) is therefore equivalent to the total entropy generation and ΔG_m being independent of time, i.e., $\Phi(t) = c_1$, $G_m^f(t) = c_2$ for $t \geq t_{eq}$, where c_1 and c_2 denote real numbers, which do not necessarily represent the extremums of $\Phi(t)$ and $G_m^f(t)$.

10.6 Semiempirical model

In a semiempirical model, an experimental relation describing conversion (or formation) of a species participating in the reaction is incorporated into an equilibrium model. This technique has been employed in some studies to reduce the calculation errors associated with the traditional equilibrium models. For example, the research team of Grace studied a pilot circulating fluidized bed coal gasifier using the traditional method of Gibbs energy minimization [21]. The gasifier had a riser with a height of 6.3 m and an inner diameter of 0.1 m. The gasification products were reported to leave the apparatus at a temperature between 700°C and 800°C. Two types of coals used for the gasification study were Highvale and Pittsburgh Seam.

The equilibrium model developed by Grace and coworkers takes 42 gaseous species and two solid species into account. As shown in Table 10.3, these 44 species involve the following five elements: carbon, hydrogen,

Table 10.3 The species considered in the equilibrium model reported in Ref. [21].

Group	Chemical formula
1	C(g), CH, CH_2, CH_3, CH_4, C_2H_2, C_2H_4, C_2H_6, C_3H_8
2	H, H_2, O, O_2, CO, CO_2, OH, H_2O, H_2O_2, HCO, HO_2
3	N, N_2, NCO, NH, NH_2, NH_3, N_2O, NO, NO_2, CN, HCN, HCNO
4	S(g), S_2(g), SO, SO_2, SO_3, COS, CS, CS_2, HS, H_2S
5	C(s), S(s)

oxygen, nitrogen, and sulfur. The gasification products then consist of a homogeneous phase comprising the gaseous species and two single-species solid phases. The reactions included in the equilibrium model are:

(a) combustion reactions

$$C+\frac{1}{2}O_2 \rightarrow CO \tag{10.23}$$

$$CO+\frac{1}{2}O_2 \rightarrow CO_2 \tag{10.24}$$

(b) carbon-steam and Boudouard reactions

$$C+H_2O \rightarrow CO+H_2 \tag{10.25}$$

$$C+CO_2 \rightarrow 2CO \tag{10.26}$$

(c) methanation

$$C+2H_2 \rightarrow CH_4 \tag{10.27}$$

(d) water-gas shift reaction

$$CO+H_2O \rightarrow CO_2+H_2 \tag{10.28}$$

The predicted best-fit equilibrium temperature substantially deviates from the actual operation temperature. The modification made by the authors was to incorporate the actual amount of carbon converted into the products, whereas the other elements were assumed to completely convert. This effect is accounted for by a fractional conversion factor, β, which can be determined by the following correlation obtained from coal gasification tests.

$$\beta=0.0647\alpha T^{0.3}+0.465 \tag{10.29}$$

for $\alpha=0.3-0.5$, $T=1000-1200$ K.

A comparison between the measured and predicted molar contents of carbon dioxide and carbon monoxide plotted against the air ratio is displayed in Fig. 10.7. The measured data are shown with symbols. The dashed lines represent the prediction of the traditional equilibrium model based on Gibbs function minimization. The solid lines denote the predictions of the modified model, which takes into account the actual conversion of carbon through Eq. (10.29). Also, the air ratio is defined as the ratio of the actual air flow to the stochiometric air required for complete combustion. Fig. 10.8 depicts a comparison between the predictions of the two models and the experiments for hydrogen and methane concentrations.

The modified model successfully captures the experimentally measured molar contents of CO, CO_2, and H_2. However, its prediction of CH_4 content is significantly lower than the measured data. The authors attributed the large concentration of methane to the pyrolysis reactions and quoted that the char gasification may be achieved to a limited extent in a pilot-scale gasifier. Furthermore, the heating value of the syngas computed by the traditional equilibrium model is reported to be substantially higher than the experiments, whereas the modified model predictions are in a good agreement with the measured values of the heating value.

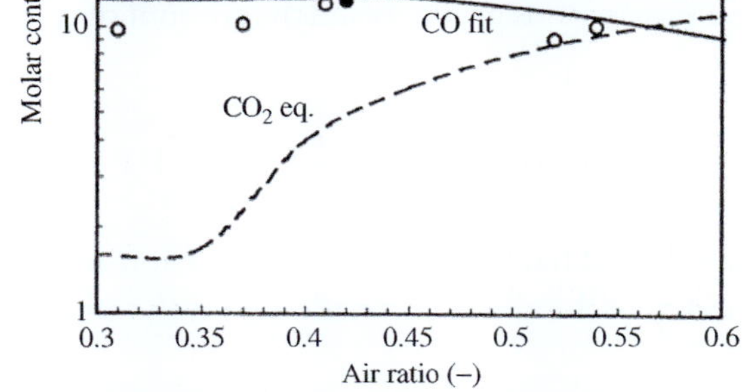

Fig. 10.7 Comparison of the measured and predicted CO and CO_2 contents using the traditional equilibrium model *(dashed lines)* and the modified model *(solid lines)* at various air ratios. Data are obtained for Highvale coal gasified at 1100 K and 155 kPa. *(From X. Li, J.R. Grace, A.P. Watkinson, C.J. Lim, A. Ergüdenler, Equilibrium modeling of gasification: a free energy minimization approach and its application to a circulating fluidized bed coal gasifier, Fuel 80 (2001) 195–207.)*

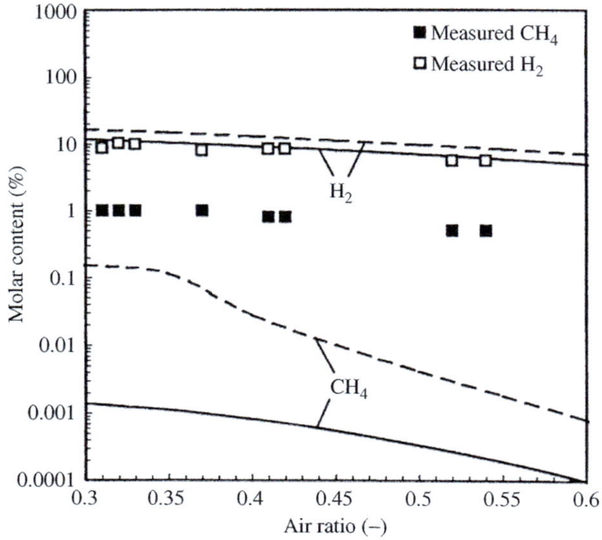

Fig. 10.8 Comparison of the measured and predicted H_2 and CH_4 contents using the traditional equilibrium model *(dashed lines)* and the modified model *(solid lines)* at various air ratios. Data are obtained for Highvale coal gasified at 1100 K and 155 kPa. *(From X. Li, J.R. Grace, A.P. Watkinson, C.J. Lim, A. Ergüdenler, Equilibrium modeling of gasification: a free energy minimization approach and its application to a circulating fluidized bed coal gasifier, Fuel 80 (2001) 195–207.)*

A few years later, the same research group extended the modified model to biomass gasification [14] by including another empirical parameter in addition to that given by Eq. (10.29) to rectify the issue of underprediction of methane concentration encountered in the earlier work [21]. A correlation is proposed in Ref. [14], which accounts for the actual conversion of carbon into gaseous products and the amount of carbon consumed to produce methane.

$$\beta_C = 0.25 + 0.75 \exp\left(-\frac{a}{0.23}\right) - 0.11(1-a) \tag{10.30}$$

where β_C denotes the net fraction of carbon conversion and a designates the air ratio. The validity of the above correlation is for the air ratio being in the range 0.21–0.54.

The conversion factor of hydrogen is determined by the following relation.

$$\beta_H = 1 - 0.44(1-a)\frac{n_C}{n_H} \tag{10.31}$$

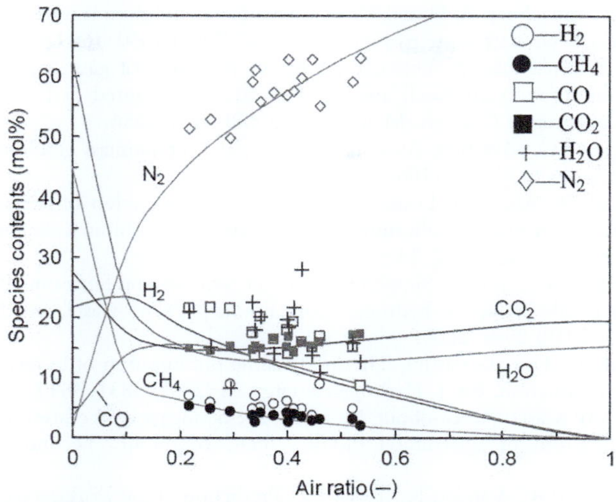

Fig. 10.9 Comparison of the measured and predicted syngas molar composition using the modified model at various air ratios. Data are obtained for various feedstock types gasified at 1100 K. *(From X. Li, J.R. Grace, A.P. Watkinson, C.J. Lim, A. Ergudenler, Biomass gasification in a circulating fluidized bed, Biomass Bioenergy 26 (2004) 171–193.)*

where n_C and n_H denote the moles of carbon and hydrogen, respectively.

In this model, it is assumed that all elements, except carbon and hydrogen, are completely converted. The predictability of the modified model is examined using a set of biomass gasification experiments obtained from a pilot-scale circulating fluidized bed gasifier. Fig. 10.9 provides a comparison between the measured and predicted dry-gas molar contents of the main species (except water) of the gasification products. The modified model predicts the CH_4 content with a good accuracy, but the predicted concentrations of hydrogen are notably higher than the experiments. Furthermore, the modified model captures the measured CO_2 contents very well, whereas it underpredicts the CO concentrations with a relative difference of 20%–25%. The authors stated that the modified model provides better predictions than the traditional Gibbs energy minimization model.

References

[1] R. Bain, Material and energy balances for methanol from biomass using biomass gasifiers, National Renewable Energy Laboratory, 1992. Report No: TP-510e17098.
[2] X. Ku, T. Li, T. Løvås, CFD–DEM simulation of biomass gasification with steam in a fluidized bed reactor, Chem. Eng. Sci. 122 (2015) 270–283.
[3] C. Di Blasi, Dynamic behaviour of stratified downdraft gasifiers, Chem. Eng. Sci. 55 (2000) 2931–2944.

[4] S. Jarungthammachote, A. Dutta, Thermodynamic equilibrium model and second law analysis of a downdraft waste gasifier, Energy 32 (2007) 1660–1669.
[5] S. Jarungthammachote, A. Dutta, Equilibrium modeling of gasification: Gibbs free energy minimization approach and its application to spouted bed and spout-fluid bed gasifiers, Energy Convers. Manage. 49 (2008) 1345–1356.
[6] M. Ruggiero, G. Manfrida, An equilibrium model for biomass gasification process, Renew. Energy 16 (1999) 1106–1109.
[7] Z.A. Zainal, R. Ali, C.H. Lean, K.N. Seetharamu, Prediction of performance of a downdraft gasifier using equilibrium modeling for different biomass materials, Energy Convers. Manage. 42 (2001) 1499–1515.
[8] M.F. Renkel, N. Lummen, Supplying hydrogen vehicles and ferries in Western Norway with locally produced hydrogen from municipal solid waste, Int. J. Hydrogen Energy 43 (2018) 2585–2600.
[9] M.R. Mahishi, D.Y. Goswami, Thermodynamic optimization of biomass gasifier for hydrogen production, Int. J. Hydrogen Energy 32 (2007) 3831–3840.
[10] S. Fournel, B. Marcos, S. Godbout, M. Heitz, Predicting gaseous emissions from small-scale combustion of agricultural biomass fuels, Bioresour. Technol. 179 (2015) 165–172.
[11] C.R. Altafini, P.R. Wander, R.M. Barreto, Prediction of the working parameters of a wood waste gasifier through an equilibrium model, Energy Convers. Manage. 44 (2003) 2763–2777.
[12] M. Baratieri, P. Baggio, L. Fiori, M. Grigiante, Biomass as an energy source: thermodynamic constraints on the performance of the conversion process, Bioresour. Technol. 99 (2008) 7063–7073.
[13] B. Acharya, A. Dutta, P. Basu, An investigation into steam gasification of biomass for hydrogen enriched gas production in presence of CaO, Int. J. Hydrogen Energy 35 (2010) 1582–1589.
[14] X. Li, J.R. Grace, A.P. Watkinson, C.J. Lim, A. Ergudenler, Biomass gasification in a circulating fluidized bed, Biomass Bioenergy 26 (2004) 171–193.
[15] S. Sharma, P.N. Sheth, Air–steam biomass gasification: experiments, modeling and simulation, Energy Convers. Manage. 110 (2016) 307–318.
[16] W.H. Chen, M.R. Lin, T.L. Jiang, M.H. Chen, Modeling and simulation of hydrogen generation from high temperature and low-temperature water gas shift reactions, Int. J. Hydrogen Energy 33 (2008) 6644–6656.
[17] P.J. Dauenhauer, J.R. Salge, L.D. Schmidt, Renewable hydrogen by autothermal steam reforming of volatile carbohydrates, J. Catal. 244 (2006) 238–247.
[18] J.W. Gibbs, On the equilibrium of heterogeneous substances, in: Transactions of the Connecticut Academy of Arts and Sciences, vol. 3, 1875, p. 108.
[19] J. M. Smith, H. C. Van Ness, M. M. Abbott, M. T. Swihart, Introduction to Chemical Engineering Thermodynamics, eighth ed., McGraw-Hill: New York.
[20] A. Gomez-Barea, B. Leckner, Modeling of biomass gasification in fluidized bed, Prog. Energy Combust. Sci. 36 (2010) 444–509.
[21] X. Li, J.R. Grace, A.P. Watkinson, C.J. Lim, A. Ergüdenler, Equilibrium modeling of gasification: a free energy minimization approach and its application to a circulating fluidized bed coal gasifier, Fuel 80 (2001) 195–207.

CHAPTER ELEVEN

Exergy

11.1 Introduction

The mechanical theory of heat, as stated by Clausius in his last memoir [1], rests on the equivalence of heat and work (first law) and the equivalence of transformations (second law). A combination of the two laws leads to a relationship between the work and the uncompensated transformation (i.e., entropy production). It was indeed Clausius who initiated utilization of the combined first and second laws. Concepts like availability and exergy (equivalent in principle) are deduced from the combined laws. They refer to the maximum work extractable from a source of heat energy, e.g., thermal reservoir, a hot stream, fuel combustion. Often, exergy-based analysis is erroneously attributed to the second law only. In this chapter, we show where exergy is originated from and that conclusions drawn from both exergy and entropy analyses should in principle be identical.

11.2 Thermal exergy

Fig. 11.1 depicts schematically an engine that operates in a closed cycle while communicating with $n+1$ thermal reservoirs. The first and second laws may analytically be expressed as

$$W_{net} = \sum_{i=0}^{n} Q_i \tag{11.1}$$

$$\Phi = -\sum_{i=0}^{n} \frac{Q_i}{T_i} \tag{11.2}$$

where T_i denotes the temperature of the ith reservoir. Also, note that the usual sign convention is applied, i.e., the heat imparted to the engine is positive whereas the heat rejected by the engine is negative.

Entropy Analysis in Thermal Engineering Systems
https://doi.org/10.1016/B978-0-12-819168-2.00011-8

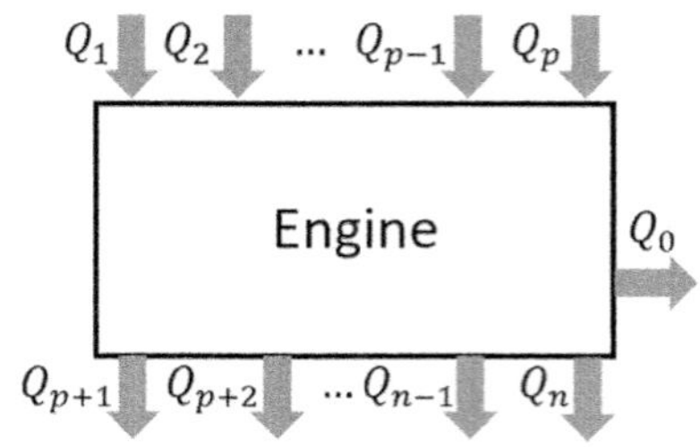

Fig. 11.1 Illustration of the concept of thermal exergy. The engine receives heat from p thermal reservoirs and rejects heat to $(n+1-p)$ thermal reservoirs. The coldest reservoir is designated with $i=0$.

Suppose that $i=0$ designates the coldest reservoir, the engine receives heat from p reservoirs and it rejects heat to $n+1-p$ reservoirs. Eqs. (11.1) and (11.2) may be rewritten as

$$W_{net} = \sum_{j=1}^{p} Q_j - \sum_{k=p+1}^{n} Q_k - Q_0 \tag{11.3}$$

$$\Phi = \frac{Q_0}{T_0} - \sum_{j=1}^{p} \frac{Q_j}{T_j} + \sum_{k=p+1}^{n} \frac{Q_k}{T_k} \tag{11.4}$$

A combination of Eqs. (11.3) and (11.4) to eliminate Q_0 yields

$$T_0\Phi + W_{net} = \sum_{j=1}^{p} \left(1 - \frac{T_0}{T_j}\right) Q_j - \sum_{k=p+1}^{n} \left(1 - \frac{T_0}{T_k}\right) Q_k \tag{11.5}$$

The simplest case is that the engine communicates with only two thermal reservoirs ($n=1$, $p=1$) characterized by their temperatures T_1 and T_0 ($< T_1$). Eq. (11.5) then reduces to

$$W_{net} = \left(1 - \frac{T_0}{T_1}\right) Q_1 - T_0\Phi \tag{11.6}$$

If the engine receives heat from a single reservoir and rejects heat to two low-temperature reservoirs ($n=2$, $p=1$), Eq. (11.6) becomes

$$W_{net} = \left(1 - \frac{T_0}{T_1}\right) Q_1 - \left(1 - \frac{T_0}{T_2}\right) Q_2 - T_0\Phi \tag{11.7}$$

It can be deduced from Eqs. (11.6) and (11.7) that minimization of entropy production may guarantee a maximum work *if* the heat transfer terms, i.e., Q_1 in Eq. (11.6), Q_1 and Q_2 in Eq. (11.7), are fixed. In general, for an engine communicating with $n+1$ thermal reservoirs, the maximum work coincides with the minimum entropy generation if all Q_j and Q_k terms in Eq. (11.5) are fixed.

Now, the thermal exergy is defined as

$$\Psi^{th} = \left(1 - \frac{T_0}{T_i}\right) Q_i \tag{11.8}$$

It represents the maximum theoretical work extractable from a given quantity of heat Q_i in a closed cycle operating between two thermal reservoirs maintained at T_i and T_0. Given the definition of the thermal exergy, Eq. (11.5) may be expressed as

$$W_{net} = \sum_{j=1}^{p} \Psi_j^{th} - \sum_{k=p+1}^{n} \Psi_k^{th} - \Psi_{de} \tag{11.9}$$

where $\Psi_{de} = T_0 \Phi$ denotes exergy destruction. If all Q_j and Q_k are assumed to be constant, Ψ_j^{th} and Ψ_k^{th} will also be constant. Then, maximization of work output would be identical to minimization of exergy destruction or of entropy generation.

11.3 Flow exergy

Suppose that the thermal energy requirement of a heat engine is supplied from a hot stream that is initially at temperature T (Fig. 11.2A). Its temperature drops within the engine, and the exhaust stream leaving the engine at T_e is discharged to the surroundings that is at T_0 ($< T_e$). Applying the first law to the engine, one obtains

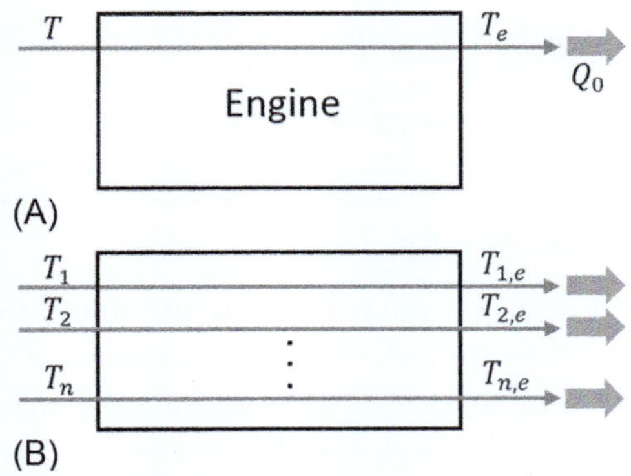

Fig. 11.2 Illustration of the concept of flow exergy. The heat engine is powered with (A) a single hot stream, (B) multiple hot streams.

$$W_{net} = H - H_e \tag{11.10}$$

The total entropy generation associated with the engine is

$$\Phi = (S_e - S) + \left(\frac{Q_e}{T_0} + S_0 - S_e\right) \tag{11.11}$$

where the first bracket denotes the change in the entropy of the hot stream within the engine, and the second bracket is the entropy generation due to cooling the exhaust stream from T_e to T_0. Also,

$$Q_e = H_e - H_0 \tag{11.12}$$

From Eqs. (11.10) and (11.12), we conclude

$$W_{net} = H - Q_e - H_0 \tag{11.13}$$

Eliminating Q_e between Eqs. (11.11) and (11.13), we get

$$T_0\Phi + W_{net} = H - H_0 - T_0(S - S_0) \tag{11.14}$$

The right-hand side of Eq. (11.14) denotes the flow exergy, so it can alternatively be expressed as

$$\Psi_{de} + W_{net} = \Psi^{fl} \tag{11.15}$$

In the reversible limit $\Psi_{de} = 0$, so Ψ^{fl} would represent the maximum theoretical work extractable from the hot stream. On the other hand, for a known initial state of the hot stream, the flow exergy in Eq. (11.15) is fixed. Thus, maximization of work would be identical to minimization of exergy destruction or of entropy generation.

In general, if the engine is operated using the thermal energy of multiple hot streams (Fig. 11.2B), Eq. (11.15) is expressed as

$$\Psi_{de} + W_{net} = \Psi_1^{fl} + \Psi_2^{fl} + \ldots + \Psi_n^{fl} \tag{11.16}$$

where Ψ_i^{fl} denotes the flow exergy of the ith stream and $i = 1, 2, \ldots, n$.

11.4 Chemical exergy

The chemical exergy is referred to the maximum theoretical work extractable from the oxidation of a fuel. We investigated the combustion of a hydrocarbon fuel in air in Chapter 8 and showed that the maximum work per unit mole of fuel, which was denoted by w_{rev}, is nearly constant. In practical power-generating systems driven by combustion of fuel in air supplied from the ambient, the fuel and air enter the system separately through two different streams. The combustion products are discharged to the atmosphere; see Fig. 11.3. The combined relation of the first and second laws for the system of Fig. 11.3 reads

$$W_{net} = \left(H_0^a + H_0^f - H_0^p\right) - T_0\left(S_0^a + S_0^f - S_0^p\right) - T_0\Phi \tag{11.17}$$

where H is the total enthalpy, superscripts a, f, and p refer to the air, fuel, and combustion products, respectively.

The maximum theoretical work of a combustion-driven engine is defined as

$$\Psi^{ch} = \left(H_0^a + H_0^f - H_0^p\right) - T_0\left(S_0^a + S_0^f - S_0^p\right) \tag{11.18}$$

where Ψ^{ch} is the chemical exergy. Eq. (11.17) may be expressed in terms of chemical exergy and exergy destruction as follows.

$$W_{net} = \Psi^{ch} - \Psi_{de} \tag{11.19}$$

Rewritten Eq. (11.19) per unit mole of fuel, we have

$$w_{net} = \psi^{ch} - \psi_{de} \tag{11.20}$$

where ψ^{ch} denotes the specific chemical exergy of the fuel (it is equivalent to w_{rev} in Chapter 8). Also,

$$\psi_{de} = T_0\text{SEG} \tag{11.21}$$

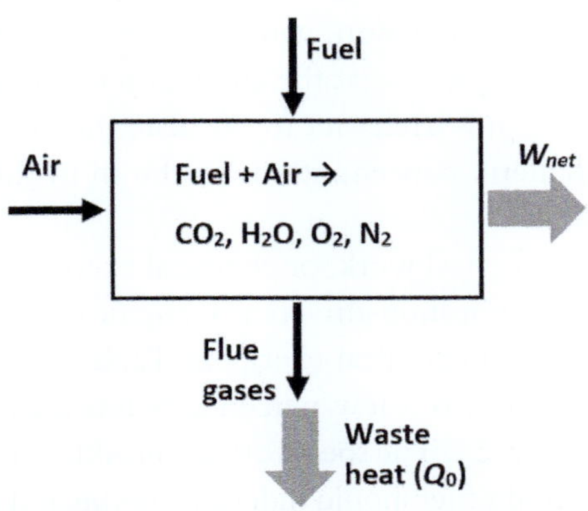

Fig. 11.3 A combustion-driven engine: illustration of the concept of chemical exergy.

where SEG is the specific entropy generation, and we may denote ψ_{de} as the specific exergy destruction (SED). It was shown in Chapter 8 that the minimization of SEG is identical to the maximization of thermal efficiency in combustion power plants.

The chemical exergy of fuel is traditionally determined with the entropies of the combustion gases evaluated at the fixed atmospheric concentrations. It is further assumed that combustion takes place in pure oxygen, water content of the products mixture remains in gas phase, and the combustion products behave like ideal gases. It should be noted that the entropy generation term in Eq. (11.17) accounts for the irreversibility within the system and that due to the rejection of the waste heat to the atmosphere. Also, the entropy of the combustion products at the exit of the system, S_e^p, and at the ambient temperature and pressure, S_0^p, should be determined based on the concentrations of the gaseous species in the combustion products. Furthermore, the possibility of condensation of water vapor needs to be accounted for; see Chapter 8.

Important to note is that heat engine is a device that converts *heat* into mechanical work. It is not designed to produce work through isothermal *mixing* of the flue gases with the atmospheric air, which would occur outside the boundary of the system. The chemical exergy of a fuel should in accordance with Eq. (11.18) represent the maximum possible work extractable from the thermal energy liberated through burning the fuel. On the other hand, as discussed in Chapter 4, the isothermal mixing of ideal gases is in principle equivalent to the expansion of individual gases from an initial pressure to a final pressure; a process which according to the first law, involves an amount of heat identical to the expansion work of the gases. If one desires to account for the mixing of the flue gases with the atmospheric air as an additional source of entropy generation, it would also be necessary to take the heat effect of the mixing process (at the ambient temperature) into consideration. To summarize, it is irrelevant to calculate the fuel chemical exergy based on the atmospheric concentrations of the individual gases released from combustion of fuel.

The maximum theoretical work, or chemical exergy, extractable from a unit mole of fuel in a combustion-driven heat engine (see Table 8.1) is compared with the traditional chemical exergy in Table 11.1 for several fuels. The chemical exergy, ψ^{ch}, is a few percentages less than the traditionally defined chemical exergy, ξ_{ch}. The specific work production of a power plant driven by combustion of a fuel should indeed be compared to ψ^{ch} in order to have an accurate measure of performance from a second law perspective.

Table 11.1 Comparison of ψ^{ch} with traditional chemical exergy ξ_{ch} taken from Ref. [2], for several fuels at standard temperature and pressure (298.15 K, 1 bar).

Fuel	Chemical formula	LHV (kJ/mol)	HHV (kJ/mol)	ψ^{ch} (kJ/mol)	ξ_{ch} (kJ/mol)	Δ (%)
Hydrogen	H_2	241.8	285.8	234.4	236.1	0.73
Methanol	CH_3OH	676.2	764.2	705.2	722.3	2.42
Methane	CH_4	802.5	890.6	814.8	831.7	2.07
Ethanol	C_2H_5OH	1278	1410	1330	1364	2.56
Acetylene	C_2H_2	1257	1301	1236	1266	2.43
Ethylene	C_2H_4	1323	1411	1329	1361	2.41
Ethane	C_2H_6	1429	1561	1464	1496	2.19
Propane	C_3H_8	2043	2219	2102	2154	2.47
Butane	C_4H_{10}	2657	2877	2740	2806	2.41
Pentane	C_5H_{12}	3272	3536	3378	3463	2.52
Octane	C_8H_{18}	5116	5512	5294	5413	2.25

11.5 A simple relation for chemical exergy

As discussed in Chapter 8, the maximum theoretical work (chemical exergy) obtainable from a unit mole of fuel represented by $C_xH_yO_z$ obeys

$$\psi^{ch} = \mathcal{C} + \mathcal{F}(\Lambda) \tag{11.22}$$

where the magnitude of function $\mathcal{F}(\Lambda)$ is negligible compared to $\mathcal{C}$. Also,

$$\mathcal{C} = HHV + T_0\mathcal{C}_1 \tag{11.23}$$

$$\mathcal{C}_1 = \left(\frac{y}{2}\right)s_{H_2O_{(l)},0} + xs_{CO_2,0} + \left(-x - \frac{y}{4} + \frac{z}{2}\right)s_{O_2,0} - s_{f,0} \tag{11.24}$$

where HHV is the higher heating value, and subscript 0 refers to 298.15 K and 1 bar.

A simple relation is now derived for ψ^{ch} using the entropies and enthalpies of water, carbon dioxide, and oxygen at standard temperature and pressure taken from the NIST database [3]. The objective is to find an equation for ψ^{ch} in terms of the elemental composition of fuel. Using the properties given in Table 11.2 in Eq. (11.24), one obtains

$$\mathcal{C}_1 = \left(\frac{y}{2}\right)(69.95) + x(213.79) + \left(-x - \frac{y}{4} + \frac{z}{2}\right)(205.15) - s_{f,0}\ (\text{J/mol K}) \tag{11.25}$$

Table 11.2 The enthalpy and entropy values of water, oxygen, and carbon dioxide at 298.15 K, 1 bar.

Substance	h (kJ/mol)	s (J/mol K)
Water (liquid)	−285.8	69.95
Water (vapor)	−241.8	188.8
Oxygen	–	205.15
Carbon dioxide	−393.5	213.79

On the other hand, the higher heating value may be represented as

$$\begin{aligned} HHV &= h_{f,0} - x h_{CO_2,0} - \left(\frac{y}{2}\right) h_{H_2O(l),0} \\ &= h_{f,0} + x(393.5) + \left(\frac{y}{2}\right)(285.8)(\text{kJ/mol}) \end{aligned} \tag{11.26}$$

Substituting Eqs. (11.25) and (11.26) into Eq. (11.23) and grouping the similar terms allows one to obtain the following expression for the chemical exergy.

$$\psi^{ch} = x(396.08) + y(138.04) + z(30.58) + \left(\tilde{h}_{f,0} - T_0 \tilde{s}_{f,0}\right) \tag{11.27}$$

where ψ^{ch} has units of kJ/mol, $\tilde{h}_{f,0}$ is the formation enthalpy of the fuel in kJ/mol, $\tilde{s}_{f,0}$ is the standard entropy of the fuel in kJ/mol K, and $T_0 = 298.15$ K.

Dividing Eq. (11.27) by the molecular weight of the fuel leads to an alternative relation for the chemical exergy of a hydrocarbon fuel in units of kJ/g.

$$\psi^{ch} = 31.1[\text{C}] + 136.1[\text{H}] + 1.91 + \left(h_{f,0} - T_0 s_{f,0}\right) \tag{11.28}$$

where [C] and [H] denote the carbon and hydrogen contents per unit mass of fuel, respectively.

Another useful correlation can be obtained for the chemical exergy as a function of the heating value of fuel using the data of Table 11.1. Hence,

$$\psi^{ch} = 0.967 HHV - 38.0 \tag{11.29}$$

Note that Eq. (11.29) is obtained using a limited number of data points. However, Eqs. (11.27) and (11.28) are derived from basic thermodynamic principles and by using the empirical data of NIST [3]. So, they can be applied for estimating the chemical exergy of any fuel with a chemical formula of $C_xH_yO_z$.

11.6 Maximum efficiency

The maximum thermal efficiency of a heat engine is defined as the maximum theoretical work per unit thermal energy supplied to the engine. The form of expression for the engine efficiency depends on whether heat is supplied from a thermal reservoir, a hot stream, or fuel. The relations that we derived for maximum work in Sections 11.2–11.4 will now be used to derive expressions for the maximum efficiency of a heat engine.

For an engine, which receives a given quantity of heat from p thermal reservoirs while communicating with $n+1$ reservoirs (Fig. 11.1), dividing the maximum work obtained from Eq. (11.9) with $\Psi_{de}=0$ by the total amount of heat imparted to the engine yields

$$\eta_{max} = \frac{\sum_{j=1}^{p} \Psi_j^{th} - \sum_{k=p+1}^{n} \Psi_k^{th}}{\sum_{j=1}^{p} Q_j} \tag{11.30}$$

In the simplest case with two thermal reservoirs ($n=1$ and $p=1$), Eq. (11.30) reduces to the Carnot efficiency.

$$\eta_{max} = \frac{\Psi_1^{th}}{Q_1} = \frac{\left(1-\frac{T_0}{T_1}\right)Q_1}{Q_1} = 1-\frac{T_0}{T_1} \tag{11.31}$$

If the engine, as another example, receives a given quantity of $Q_1 + Q_2$ heat from two thermal reservoirs ($p=2$) and rejects heat to only one reservoir ($n=2$), we have

$$\eta_{max} = \frac{\sum_{j=1}^{2} \Psi_j^{th}}{\sum_{j=1}^{2} Q_j} = \frac{\left(1-\frac{T_0}{T_1}\right)Q_1 + \left(1-\frac{T_0}{T_2}\right)Q_2}{Q_1 + Q_2} \tag{11.32}$$

The expression for the maximum efficiency of an engine that receives heat from a hot stream obeys

$$\eta_{max} = \frac{\Psi^{fl}}{H - H_0} = 1 - \frac{T_0(S - S_0)}{H - H_0} \tag{11.33}$$

To establish an expression for the maximum efficiency of an engine driven by fuel combustion, we note that the thermal energy input per unit mole of fuel is $HHV - n_v\mathcal{L}_0^w$ (see Eq. 8.11) and the maximum work is the chemical exergy of the fuel given by Eq. (11.22). Hence,

$$\eta_{max} = \frac{\mathcal{C} + \mathcal{F}(\Lambda)}{HHV - n_v\mathcal{L}_0^w} \tag{11.34}$$

An important conclusion from Eqs. (11.30)–(11.34) is that from a second law perspective the upper limit of a heat engine efficiency may differ from one design to another. It would therefore be inappropriate to use the Carnot efficiency as the upper limit of efficiency for all heat engines. For further clarification, the maximum efficiency of a combustion-driven engine computed using Eq. (11.34) at stoichiometric combustion for several fuels is provided in Table 11.3. The efficiency of a Carnot cycle, Eq. (11.31), operating between $T_0 = 298.15$ K and the adiabatic flame temperature is also given for comparison. It can be inferred from Table 11.3 that for all fuels (except hydrogen), the maximum theoretical efficiency of the actual system is significantly higher than the efficiency of a Carnot engine operating between the same high and low temperatures. This result highlights the shortcoming of using the Carnot efficiency as the upper limit of combustion-driven

Table 11.3 Comparison of the maximum efficiency of a combustion-driven engine with the corresponding Carnot efficiency.

Fuel	Adiabatic flame temperature (K)	Maximum efficiency	Carnot efficiency
Hydrogen	2519	82.8	88.2
Methanol	2329	93.4	87.2
Methane	2325	92.8	87.2
Ethanol	2351	95.7	87.3
Acetylene	2909	96.2	89.8
Ethylene	2564	95.5	88.4
Ethane	2379	95.1	87.5
Propane	2391	96.1	87.5
Butane	2397	96.6	87.6
Pentane	2401	96.9	87.6
Hexane	2404	97.1	87.6
Octane	2407	97.5	87.6

power-generating systems. It reveals that the efficiency of a Carnot cycle constrained to operate between the same highest and lowest temperatures of a combustion-driven cycle is not an accurate efficiency bound.

Based on this argument, it would be inappropriate to define the second law efficiency of a heat engine as the ratio of the actual thermal efficiency (η_a) to the efficiency of the corresponding Carnot engine. It should rather be defined as the ratio of η_a to η_{max}. Hence,

$$\eta_{SL} = \frac{\eta_a}{\eta_{max}} \tag{11.35}$$

The second law efficiency defined in Eq. (11.35) accounts for the heat source type because as discussed, the maximum theoretical work production of a heat engine may vary depending on whether the thermal energy requirement of the engine is supplied from thermal reservoir(s), hot stream(s), or fuel.

11.7 Minimum exhaust temperature

According to the second law, a portion of the thermal energy supplied to a heat engine must be rejected to the surroundings, even at fully reversible operation where entropy generation is zero. From this principle, we may define a minimum exhaust temperature for an engine that is run by a hot stream or fuel combustion. In this case, the amount of heat rejected by the engine is obtained using Eq. (11.12). If the heat is supplied by a hot stream (Fig. 11.2A), the heat rejected at the reversible limit can also be determined by

$$Q_e^{rev} = (H - H_0) - \Psi^{fl} = T_0(S - S_0) \tag{11.36}$$

Comparing Eq. (11.12) with Eq. (11.36), one obtains

$$H_e^{rev} - H_0 = T_0(S - S_0) \tag{11.37}$$

Notice that H_e denotes the enthalpy of the hot stream at the engine exit, and S is the entropy of the stream at the inlet of the engine. Assume that the hot steam is an ideal gas with a uniform pressure. The minimum exhaust temperature or the exhaust temperature at reversible operation may be determined by

$$T_e^{rev} = T_0\left(1 + \ln\frac{T}{T_0}\right) \tag{11.38}$$

On the other hand, if the engine is run by fuel combustion, we find the following relation.

$$H_e^{rev} = H_0 + \left(\frac{1}{\eta_{max}} - 1\right)\Psi^{ch} \tag{11.39}$$

Here, H_e^{rev} denotes the enthalpy of the combustion products calculated at T_e^{rev}, which designates the minimum theoretical temperature of the combustion products exiting the engine. For instance, if methane is used as a fuel, the minimum exhaust temperature according to Eq. (11.39) is 313.8 K. An exit temperature lower than T_e^{rev} would be in violation of the laws of thermodynamics.

11.8 Final notes

Before closing this final chapter, there remains to make few remarks about the limitation of the second law and the application of entropy-based analysis.

11.8.1 Entropy vs exergy

In principle, there should not be any difference between the design results obtained from entropy and exergy analyses. It was repeatedly concluded in Sections 11.2–11.4 that the exergy destruction is equivalent to the total entropy generation multiplied by the temperature of the coldest reservoir (Section 11.2) or of the cold reservoir (Sections 11.3 and 11.4). So, minimization of exergy destruction is identical to minimization of total entropy generation.

The simplest explanation for the coincidence of the optimum of the two thermodynamic functions would, perhaps, be: exergy is a consequence of the first and second laws, but because the first law is an expression of the conservation of energy, an exergy-based analysis applied to a steady-state process carries a function whose net value is zero. For a system undergoing a steady-state process, the first law may be represented with a single function as $F(E_i) = 0$, where E_i denotes all forms of energy that may appear in a first law equation.

It would then be a matter of preference to employ an entropy- or exergy-based analysis when modeling the thermodynamic performance of an energy conversion system. A thermodynamic model is primarily built upon the mass conservation principle and the first law. If one decides to also include an exergy analysis in the thermodynamic model, it would be at the cost of incorporating the function $F(E_i)$ in the model whose effect would eventually be null. Thus, it would comparably be less cumbersome to simply use entropy analysis which would yield a relatively less complex thermodynamic model than incorporating exergy analysis.

11.8.2 Limitation of the second law

Thermodynamics is the science of heat and work. It was referred by its founders as the *Mechanical Theory of Heat*. The first law was founded on the observations that heat can be produced by spending work, and work may be generated by consumption of heat. The second law is based on the impossibility of complete conversion of heat to work in an engine, or impossibility of spontaneous transference of heat from a cooler to a warmer body. The formulation of the second law, however, is based on a *special* condition, i.e., reversibility, that guarantees a production of maximum work from a given amount of heat. The condition of reversibility is characterized by zero entropy generation.

Unlike the first law that can be applied to a variety of processes with or without a presence of work, an application of the second law to a system with no work interaction is not expected to produce meaningful results. The first law is a form of the conservation of energy; a general principle whose truth is independent of the nature or kind of energy. On the other hand, the formulation of the second law rests on the presence of both heat and work. It may simply be stated as the equivalence of reversibility, zero entropy generation, and maximum work; the second law requires all these three elements at once. If this be agreed as a simple statement of the second law, the question is then what would reversibility or zero entropy generation represent in a system that does not include any work interaction externally?

Not every machine invented by mankind is a heat engine, a refrigerator, or a heat pump. There are many devices that are designed to fulfill other objectives where a production or consumption of work (power) is of no or minor importance. An example is heat exchangers whose primary task is to transfer heat from a hot fluid to a cold fluid. Once the hot fluid is cooled down to a desired temperature and, at the same time, the cold fluid is heated up to a preset temperature, it may be said that the heat exchanger has fulfilled its objective. In addition to exchanging heat, the two fluids may experience retardation of motion due to, for example, friction, whereby heat is generated due to the frictional work. Thus, the entropy generation due to pressure drop is directly related to an actual loss of useful work, whereas that associated with heat transfer between the two fluids is of no practical concern. Indeed, in a heat exchanger, the former should be minimized while the latter is enhanced.

11.8.3 Application area of entropy analysis

The applicability of entropy-based analysis is limited. It may lead to *meaningful* results in thermal power–generating systems and subject to certain design constraints. For instance, as discussed in Chapter 8, in a power plant driven by fuel combustion, the thermal efficiency is inversely proportional to the *specific entropy generation* (SEG) defined as the ratio of the total entropy generation rate per unit flowrate of the fuel burnt. For an engine operating between two thermal reservoirs, the maximum efficiency design would be identical to the minimum entropy generation design if either the heat input or the work output is fixed. Furthermore, it must be noted that mere calculation of the rate of entropy generation (or exergy destruction) is insufficient. Such calculations should enable designer to identify ways that would boost the performance of a power-generating system, which would otherwise be of no practical value.

Finally, the journey of this book comes to an end with an astonishing note of Carnot as his final words in *Réflexions sur la Puissance Motrice du Feu*, who about two centuries ago discouraged any attempt to mere achieving maximum theoretical power in heat engines [4,5].

> *We should not expect ever to utilize in practice all the motive power of combustibles. The attempts made to attain this result would be far more hurtful than useful if they caused other important considerations to be neglected. The economy of the combustible is only one of the conditions to be fulfilled in heat-engines. In many cases it is only secondary. It should often give precedence to safety, to strength, to the durability of the engine, to the small space which it must occupy, to small cost of installation, etc.*

References

[1] R. Clausius, Ueber verschiedene für die Anwendung bequeme Formen der Hauptgleichungen der mechanischen Wärmetheorie, Ann. Phys. (1865). Bd. cxxv, 353.
[2] M.J. Moran, H.N. Shapiro, D.D. Boettner, M.B. Bailey, Fundamentals of Engineering Thermodynamics, ninth ed., Wiley, Hoboken, NJ, 2018.
[3] NIST Chemistry WebBook, SRD 69. National Institute of Standards and Technology, https://webbook.nist.gov/chemistry/form-ser/.

[4] S. Carnot, Réflexions sur la Puissance Motrice du Feu et Sur Les Machines Propres à Développer Cette Puissance, Chez Bachelier, Paris, 1824, p. 117.
[5] S. Carnot, R.H. Thurston (Ed.), Reflections on the Motive Power of Heat, second ed., Wiley, New York, 1897, p. 126.

Nomenclature

a_j	stochiometric coefficient of species *j*, Eq. (10.15)
c_p	specific heat at constant pressure
c_v	specific heat at constant volume
C	concentration
$\dot{C}$	$=\dot{m}c_p$
CR	compression ratio
E	expansion coefficient, Eqs. (2.11), (2.12), (2.13), (3.7)
E	energy
$\dot{E}$	rate of energy
$f()$	function
g_m	$=h-T_s s$
G	Gibbs function
G_m	$=ng_m$, Eq. (10.10)
h	specific enthalpy
h_f	specific enthalpy of formation
H	enthalpy
HV	heating value
I	electric current
$\mathcal{L}_0^w$	evaporation enthalpy of water
K	thermal conductance
m	mass
$\dot{m}$	mass flowrate
n	number of moles
$\dot{n}$	molar flowrate
N_e	number of electrons
p	pressure
PR	pressure ratio
q	heat per unit mass (mole)
Q	heat
$\dot{Q}$	heat rate
r	reaction rate
r_K	$=K_h/K_l$, Eq. (6.11)
r_T	$=T_H/T_L$, Eq. (6.11)
R	Universal gas constant
R_g	gas constant
s	specific entropy
S	entropy
SEG	specific entropy generation
t	temperature in Centigrade degree, Chapter 2
t	time
T	temperature in Kelvin
TIT	turbine inlet temperature
T_R	ratio of the highest-to-lowest temperatures, Eq. (5.11)

U internal energy
U_a air utilization factor, Chapter 9
U_f fuel utilization factor, Chapter 9
V volume
V voltage, Eq. (9.23)
w specific work
W work
$\dot{W}$ power
X conversion
y_i mole fraction of component i

Greek letters

α expansion coefficient, Eq. (2.10)
α parameter defined in Eq. (7.12)
β parameter defined in Eq. (7.13)
β fractional conversion factor, Eq. (10.29)
Δ quantitative variation/difference
γ $=c_p/c_v$
ϵ heat exchanger effectiveness, Eq. (8.49)
ε negligible real number
η efficiency
η_{SL} second law efficiency
κ reaction constant, Eq. (10.22)
Λ stochiometric coefficient
μ Carnot's function, Eqs. (2.11) and (2.13)
ξ reaction advancement
π $=T_H/T_L$, Eq. (7.15)
ρ density
ϕ fuel-air equivalence ratio
Φ entropy generation
$\dot{\Phi}$ entropy generation rate
ψ specific exergy
Ψ exergy

Subscripts

0 Reference state, 298.15 K & 1 bar
ad adiabatic
c compressor

com	compression
com	combustor, Eq. (8.42)
cp	condensate pump
eq	equilibrium
exp	expansion
fc	fuel compressor
fp	feedwater pump
g	gas
in	input
H	high-temperature reservoir
L	low-temperature reservoir
max	maximum
min	minimum
mix	mixture
OC	open circuit
opt	optimum
P	products
R	reactants
rev	reversible
s	surrounding
sat	saturation
st	steam turbine
sys	system
t	turbine
th	thermal
tot	total

Superscripts

*	normalized
a	air
ch	chemical
f	final state
f	fuel, Chapter 8
fl	flow, Eq. (11.15)
FC	fuel cell
i	initial state
HF	hot fluid
p	reaction products

Appendices

Appendix A: Efficiency at maximum power

An interesting characteristic of most endo-reversible engines is the maximum power efficiency; i.e.,

$$\eta = 1 - \sqrt{\frac{T_L}{T_H}} \tag{A.1}$$

For many years after Eq. (A.1) appeared in Curzon and Ahlborn (1975)'s paper, the efficiency in Eq. (A.1) carried the subscripts "CA" denoting the initials of the authors last names. Later, it turned out that Eq. (A.1) had been presented about two decades earlier by Chambadal and Novikov. There is however evidence [1] that the expression for the efficiency at maximum power had already been derived by James Henry Cotterill, Professor of Applied Mechanics in the Royal Naval College, in the late 19th century. The maximum power efficiency can be found in the second edition of the Cotterill's textbook [1], Chapter IV, pages 100–102. The author was unsuccessful in locating the first edition of the book published in 1877 to confirm whether Eq. (A.1) was first given in the earliest edition. Nevertheless, important to remember is that it would be inappropriate to refer Eq. (A.1) as Chambadal-Novikov-Curzon-Ahlborn efficiency.

Appendix B: Effect of fuel type on SEG

To examine the effect of the fuel type on SEG, the efficiency and the specific entropy generation are calculated and compared for the gas turbine cycle studied in Chapter 8 operating on hydrogen, propane, methanol, and ethanol. At a given pressure ratio and TIT, the specific entropy generation is quantitatively different depending on the type of the fuel burnt. The highest and lowest values of SEG (measured in J/mol K) are obtained for propane (C_3H_8) and hydrogen, respectively. A fuel with a greater heating value would yield a higher SEG. The heating values of the fuels and the minimum specific entropy generation of the cycle at TIT = 1100 K are compared in Table B.1. Note that if the calculations are performed on mass basis the minimum SEG would still correlate with the heating value. In this case,

Table B.1 Comparison of the heating value and the minimum specific entropy generation of a gas turbine cycle (TIT = 1100 K) for five different fuels.

Fuel	HV (kJ/mol)	Minimum SEG (J/mol K)
H_2	241.8	536
CH_3OH	675.9	1676
CH_4	802.3	1956
C_2H_5OH	1277.5	3187
C_3H_8	2043.9	5075

hydrogen with the highest heating value (measured in kJ/g) among the five fuels would yield the greatest specific entropy generation, whereas methanol with the least heating value would lead to the lowest SEG.

Appendix C: Determination of ξ at minimum G_m^f

An equation can be derived for the reaction advancement that minimizes the function G_m^f but maximizes the entropy generation, see Eq. (10.10), for a mixture of k ideal gases, i.e., Eq. (10.19). Substituting the relation

$$s_j(T,p) = s_j^0 + s_{T,j} - R\ \ln y_j^f \tag{C.1}$$

for the entropy of species j in Eq. (10.19) yields

$$G_m^f(\xi) = \sum_{j=1}^{k} \left(n_j^i + a_j\xi\right)\left[h_j(T) - T_s\left(s_j^0 + s_{T,j} - R\ \ln y_j^f - R\ \ln p\right)\right] \tag{C.2}$$

where s^0 denotes the specific entropy at the standard temperature and pressure, and s_T is the entropy change due to the difference between the temperature T and the standard temperature.

The model fraction of species j is defined as

$$y_j^f = \frac{n_j^f}{n^f} = \frac{n_j^i + a_j\xi}{n^i + a\xi} \tag{C.3}$$

where n_j^f and n^f are substituted from Eqs. (10.15) and (10.17), respectively.

Substituting Eq. (C.3) into Eq. (C.2), G_m^f can be described as a function of ξ only.

$$G_m^f(\xi) = \sum_{j=1}^{k} \left(n_j^i + a_j\xi\right) \left[h_j(T) - T_s\left(s_j^0 + s_{T,j} - R\ln\left(\frac{n_j^i + a_j\xi}{n^i + a\xi}\right) - R\ln p\right)\right] \tag{C.4}$$

At the minimum G_m^f, one must solve $dG_m^f/d\xi = 0$ to get

$$\sum_{j=1}^{k} \left\{ a_j\left[h_j(T) - T_s\left(s_j^0 + s_{T,j}\right)\right] + T_s R\ln\left(\frac{n_j^i + a_j\xi}{n^i + a\xi}\right)^{a_j} + T_s R\ln p^{a_j} \right\} = 0 \tag{C.5}$$

Upon simplification and rearrangement, we obtain

$$\prod_{j=1}^{k} \left(\frac{n_j^i + a_j\xi}{n^i + a\xi}\right)^{a_j} = \exp\left\{ -\frac{1}{T_s R} \sum_{j=1}^{k} \left\{ a_j\left[h_j(T) - T_s\left(s_j^0 + s_{T,j}\right)\right] \right\} - a\ln p \right\} \tag{C.6}$$

For a reactive system with a known initial state and composition that interacts with its surrounding that is at temperature T_s, the only unknown in Eq. (C.6) is ξ. If ξ at minimum G_m^f happens to be close to ξ_{eq} (as was seen in Fig. 10.6) the composition of the mixture at the final state may then be readily obtained from Eq. (C.3) and using ξ determined from Eq. (C.6).

Reference

[1] J.H. Cotterill, The Steam Engine Considered as a Thermodynamic Machine, A Treatise on the Thermodynamic Efficiency of Steam Engines, second ed., E & F. N. Spon, London, 1890.

Index

Note: Page numbers followed by *f* indicate figures and *t* indicate tables.

R

S

T

U

W

Z

CPI Antony Rowe
Eastbourne, UK
November 20, 2019